Code of Practice for
Project Management
for Construction and Development

ORES UNIVERSITY
Rs L.R.C.
3701/3634

108
119
120

D0244063

LIVERPOOL JOHN MOORES UNIVERSITY
Aldham Robarts L.R.C.
TEL. 051 231 3701/3634

LIVERPOOL JMU LIBRARY

3 1111 01030 2865

Code of Practice for
Project Management
for Construction and Development

Third edition

© 2002 The Chartered Institute of Building

Blackwell Publishing Ltd
Editorial Offices:
Osney Mead, Oxford OX2 0EL, UK
  Tel: +44 (0)1865 206206
Blackwell Science, Inc., 350 Main Street,
Malden, MA 02148-5018, USA
  Tel: +1 781 388 8250
Iowa State Press, a Blackwell Publishing Company,
2121 State Avenue, Ames, Iowa 50014-8300, USA
  Tel: +1 515 292 0140
Blackwell Publishing Asia Pty, 550 Swanston Street,
Carlton, Victoria 3053, Australia
  Tel: +61 (0)3 9347 0300
Blackwell Wissenschafts Verlag,
Kurfürstendamm 57, 10707 Berlin, Germany
  Tel: +49 (0)30 32 79 060

The right of the Author to be identified as the Author
of this Work has been asserted in accordance with the
Copyright, Designs and Patents Act 1988.

All rights reserved. No part of this publication may be
reproduced, stored in a retrieval system, or transmitted, in
any form or by any means, electronic, mechanical,
photocopying, recording or otherwise, except as permitted
by the UK Copyright, Designs and Patents Act 1988,
without the prior permission of the publisher.

First published 1992
Second edition 1996
Third edition 2002 by Blackwell Publishing
Reprinted 2003

Library of Congress
Cataloging-in-Publication Data
Code of practice for project management for construction
and development – 3rd ed.
    p. cm.
  "Chartered Institute of Building."
  Includes bibliographical references and index.
  ISBN 1-40510-309-4 (alk. paper)
  1. Building–Superintendence. 2. Project management.
  I. Chartered Institute of Building (Great Britain)
TH438.C626 2003
690′.068–dc21   2002035617

ISBN 1-4051-0309-4

A catalogue record for this title is available from the
British Library

Typeset and produced by Gray Publishing,
Tunbridge Wells, Kent
Printed and bound in Great Britain by
Ashford Colour Press, Gosport

For further information on Blackwell Publishing, visit
our website: www.blackwellpublishing.com

Contents

Part 1   Project management   1

Contents

Appendices

Foreword

Much has changed since the first publication of the CIOB *Code of Practice for Project Management* in 1991; nothing more so than the demands placed on the participants in the construction process.

My view of this third edition is that it is an excellent authoritative reference to the principles and practice of project management in construction and development. It will be of value to clients, project managers and all participants in the construction process as well as to educational establishments of all types. In addition, much of the information contained in the code will also be relevant to project managers operating in other commercial spheres.

Effective project management involves the assessment and management of risk. This is a strong theme that runs throughout the code, from inception to completion. Each stage in the project process, which is, as would be expected, strong in its construction theme, is described within the code which contains a broad body of knowledge brought about through the experience of the contributors. The detail contained here will be helpful throughout all the stages of a project, but particularly at inception where client involvement and pre-planning are emphasised and again at the latter stages where facilities management and occupation by the client are considered.

I strongly commend this valuable multi-institutional code of practice to all those involved in construction project management and development in the hope that a greater degree of uniformity and clarity may be achieved in this highly fragmented industry.

Sir Stuart Lipton
Chief Executive
Stanhope plc

Preface

The aim of this *Code of Practice* is to provide the clients and all the other members of the project team with a definitive strategy for the management and co-ordination of any project. The objective of the code is to define the responsibilities of all the participants involved in order to achieve the completion of the project on time, to the specifications defined by the project brief and within the budget. The client expects that effective project management will enable the project's completion, by the time when it is wanted, of a standard and quality that is required, and at a price that is competitive.

The third edition of this *Code of Practice* is a client-orientated document. The role and responsibilities are clearly defined in this revised code – whether the client is an individual, a corporate body or a development company and whatever the form of contract they choose to use. The code helps a project manager define and recommend the appropriate form of contract to meet the client's requirements and ensure that the roles and responsibilities are defined and linked to the form of contract recommended.

The code represents a cross-boundary approach to construction project management by incorporating working practices and policies from across the construction industry, encompassing the entire range of clients, architects, engineers, quantity surveyors, builders, specialist contractors and the major professional institutions.

This third edition reflects the changes in the construction practices initiated through the Latham, Egan and other reports and represents a cohesive initiative to formulate guidelines and working practices covering the development of a construction project from inception through all its stages to completion and occupation of the developed facility.

The principles of project management are the same for any size of project. Therefore, the code is equally applicable to the greater number of smaller valued projects. The code recognises that each project is unique and that the means by which it may be procured will be subject to variation.

The structure of the code mirrors generally the project management process itself. The key issues considered are under the headings of inception, feasibility, strategy, pre-construction, construction, engineering services commissioning, completion, handover and occupation, and post-completion review/project close-out. Each chapter deals with a specific stage of project management and is supported by specimen forms, checklists and examples of typical documentation. It is pertinent to point out that the specimen forms, charts and checklists cannot be regarded as appropriate for universal application: they are only examples and their value must be assessed for the project in question.

Most importantly, there was a general consensus among the steering group that the *Code of Practice for Project Management for Construction and Development* is the only authoritative code for project management and no company or individual involved in construction project management should risk being without it.

Acknowledgements

In August 2001 the then President of the Chartered Institute of Building (CIOB), Bob Heathfield, invited other built environment professional bodies to participate in a review to create a third edition of the *Code of Practice for Project Management for Construction and Development*.

A working group was promptly established under the chairmanship of Derek Hammond. The objectives of the review were clearly defined and work on the review commenced immediately. A list of participants and the organisations represented is included in the book.

As with all good projects, the review required managing. In the CIOB's technical manager, Saleem Akram, the group found a manager who, through his skill and experience, was able not only to meet, but also to exceed the demands placed on him in delivering the review. Saleem has been ably assisted by Arnab Mukherjee both in the technical aspects of the review and in updating.

It is well known that construction people work well in teams, and this review has been achieved by all the participants co-operating with each other. Particular thanks must go to Sue Dennison and Ethi Oepen of Professional and Technical Development at the CIOB for bringing together all the disparate elements of the review of the code to enable the working group to build positively on the regular progress made between and during meetings.

The third edition of the code is more broadly representative than previous editions. It now includes representative contributions from built environment specialists through all the phases of a project. It has, as has its predecessors, been produced through interdisciplinary co-operation between professionals within the built environment.

On behalf of the CIOB I would like to thank all the members of the working group who have made a major contribution to the review through their time and experience. In particular, my sincere thanks go to those mentioned above whose contributions to the review have made it a real pleasure to be involved with.

The CIOB gratefully acknowledges the many organisations who have contributed their time and expertise and where applicable for allowing us to reproduce extracts of their own documentation.

Finally, I know I speak for all members of the working group in thanking Derek Hammond for his contribution in the development of the code. This is Derek's third experience of chairing the group and his knowledge, experience and leadership have been invaluable.

Chris Williams
Director
Professional and Technical Development

Third Working Group for the Revision of the Code of Practice for Project Management

F A Hammond MSc Tech CEng MICE FCIOB MASCE FCMI	–	Chairman
Martyn Best BA Dip Arch RIBA	–	Royal Institute of British Architects
Allan Howlett CEng FIStructE MICE MIHT	–	Institution of Structural Engineers
Gavin Maxwell-Hart BSc CEng FICE FIHT MCIArb	–	Institution of Civil Engineers
Roger Waterhouse MSc FCIOB FRICS MSIB FAPM	–	Royal Institution of Chartered Surveyors and Association for Project Management
Richard Biggs MSc FCIOB MAPM MCMI	–	Association for Project Management
John Campbell	–	Royal Institute of British Architects
Mary Mitchell	–	Confederation of Construction Clients
Jonathan David BSc MSLL	–	Chartered Institution of Building Services Engineers
Neil Powling FRICS DipProjMan (RICS)	–	Royal Institution of Chartered Surveyors
Brian Teale CEng MICBSE DMS		
David Trench CBE FAPM FCMI		
Professor John Bennett FRICS DSc		
Peter Taylor FRICS		
Barry Jones FCIOB		
Professor Graham Winch PhD MCIOB MAPM		
Ian Guest BEng		
Ian Caldwell BSc BArch RIBA ARIAS MIMgt		
J C B Goring MSc BSc (Hons) MCIOB MAPM		
Artin Hovsepian BSc (Hons) MCIOB MASI		
Alan Beasley		
David Woolven MSc FCIOB		
Colin Acus		
Chris Williams DipLaw DipSury FCIOB MRICS FASI		
Saleem Akram BEng MSc (CM) PE FIE MASCE MAPM MACostE	–	Secretary and Technical editor of third edition
Arnab Mukherjee BEng MSc (CM)	–	Assistant technical editor
John Douglas	–	Englemere Ltd

First and Second Working Groups of the Code of Practice for Project Management

F A Hammond MCs Tech CEng MICE FCIOB MASCE FIMgt	–	Chairman
G S Ayres FRICS FCIArb FFB	–	Royal Institution of Chartered Surveyors
R J Cecil DipArch RIBA FRSA	–	Royal Institute of British Architects
D K Doran BSc Eng DIC FCGI CEng FICE FIStructE	–	Institution of Structural Engineers
R Elliott CEng MICE	–	Institution of Civil Engineers
D S Gillingham CEng FCIBSE	–	Chartered Institution of Building Services Engineers
R J Biggs MSc FCIOB MIMgt MAPM	–	Technical editor of second edition Association for Project Management
J C B Goring BSc (Hons) MCIOB MAPM		
D P Horne FCIOB FFB FIMgt		
P K Smith FCIOB MAPM		
R A Waterhouse MSc FCIOB MIMgt MSIB MAPM		
S R Witkowski MSc (Eng)	–	Technical editor of first edition
P B Cullimore FCIOB ARICS MASI MIMgt	–	Secretary

For the second edition of the code changes were made to the working group which included

L J D Arnold FCIOB		
P Lord AA Dipl (Hons) RIBA PPCSD FIMgt	–	Royal Institute of British Architects (replacing R J Cecil deceased)
N P Powling Dip BE FRICS Dip Proj Man (RICS)	–	Royal Institution of Chartered Surveyors
P L Watkins MCIOB MAPM	–	Association for Project Managers

List of tables and figures

List of Tables

List of Figures

From the President

During his presidency, my predecessor, Bob Heathfield, invited the practitioners of the built environment professions to participate in the revision of the *Code of Practice for Project Management for Construction and Development*.

I am very pleased during this, my year of presidency, to endorse the work of the review group under the chairmanship of Derek Hammond on behalf of the Chartered Institute of Building.

I commend the work and the third edition and confirm it is a work of excellence.

Stuart Henderson
President
Chartered Institute of Building

Introduction

Project management

Project management is the professional discipline which separates the management function of a project from the design and execution functions. Management and design may still be combined on smaller projects and be performed by the leader of the design team. For larger or more complex projects the need for separate management has resulted in the evolution of project management.

Project management has a long history, but in its modern form its use for construction only extends back for as little as 30–40 years. Much of the earlier codification of the principles and practices of project management was developed in the United States, although the Chartered Institute of Building published its seminal work on the subject in 1979.

Project management may be defined as 'the overall planning, co-ordination and control of a project from inception to completion aimed at meeting a client's requirements in order to produce a functionally and financially viable project that will be completed on time within authorised cost and to the required quality standards.'

This Code of Practice is the authoritative guide and reference to the principles and practice of project management in construction and development. It will be of value to clients, project management practices and educational establishments/students and to the construction industry in general. Much of the information contained in the code will also be relevant to project management operating in other commercial spheres.

Role

Project management has a strong tradition in the construction industry and is widely used on projects of all sizes and complexity. Even so, many projects do not meet their required performance standards or are delivered late/over budget. These issues can be directly addressed by raising the standards of project management within the construction industry and more specifically improving the skills of project managers.

There has been a recent groundswell within the construction industry for improvement in all areas of its activities. Many of these improvements are highlighted in the 1998 Egan Report and in the ongoing work of cross-industry representative bodies such as the Construction Industry Council (CIC), Construction Industry Research and Information Association (CIRIA), Construction Best Practice Programme (CBPP) and Strategic Forum for Construction (SFC). Project management is a process which runs throughout the construction life cycle and so touches all associated activities.

Purpose of construction project management

The purpose of project management in the construction industry is to add significant and specific value to the process of delivering construction projects. This is achieved by the systematic application of a set of generic project-orientated management principles throughout the life of a project. Some of these techniques have been tailored to the sector requirements unique to the construction industry.

The function of project management is applicable to all projects. However, on smaller or less complex projects the role may well be combined with another discipline, e.g. leader of the design team. The value added to the project by project management is unique: no other process or method can add similar value, either qualitatively or quantitatively.

Structure of project management

Construction and development projects involve the co-ordinated actions of many different professionals and specialists to achieve defined objectives. The task of project management is to bring the professionals and specialists into the project team at the right time to enable them to make their best possible contribution, efficiently.

The professionals and specialists bring knowledge and experience that contribute to decisions, which are embodied in the project information. The different bodies of knowledge and experience all have the potential to make important contributions to decisions at every stage of projects. In construction and development projects there are far too many professionals and specialists involved for it to be practical to bring them all together at every stage. This creates a dilemma because ignoring key bodies of knowledge and experience at any stage may lead to major problems and additional costs for everyone.

The practical way to resolve this dilemma is to structure carefully the way the professionals and specialists bring their knowledge and experience into the project team. The most effective general structure is formed by the eight project stages used in this code's description of project management. In many projects there is a body of knowledge and experience in the client organisation. This also has to be tapped at the right time and blended with the professional and specialists' expertise.

Each stage in the project process is dominated by the broad body of knowledge and experience that is reflected in the stage name. As described above, essential features of that knowledge and experience need to be taken into account in earlier stages if the best overall outcome is to be achieved. The way the professionals and specialists who own that knowledge and experience are brought into the project team at these earlier stages is one issue that needs to be decided during the Strategy Stage.

The results of each stage influence later stages and it may be necessary to involve the professionals and specialists who undertook earlier stages to explain or review their decisions. Again the way they are employed should be decided in principle during the Strategy Stage.

Each stage relates to specific key decisions. Consequently, many Project Teams hold a key decision meeting at the end of each stage to confirm that the necessary actions and decisions have been taken and the project can therefore begin the next stage. There is a virtue in producing a consolidated document at the end of each page that is approved by the Client Body before proceeding to the next stage. This acts as a reference or peg in the sand as well as acting as a vehicle for widespread ownership.

Projects begin with the Inception Stage that results from business decisions by the client which suggest a new construction or development project may be required. Essentially, the Inception Stage consists of commissioning a project manager to undertake the next stage, which is to test the feasibility of the project. The Feasibility Stage is a crucial stage in which all kinds of professionals and specialists may be required to bring many kinds of knowledge and experience into a broad-ranging evaluation of feasibility. It establishes the broad objectives for the project and so exerts an influence throughout subsequent stages.

1st

The next stage is the Strategy Stage which begins when the project manager is commissioned to lead the project team to undertake the project. This stage requires the project's objectives, an overall strategy and the selection of key team members to be considered in a highly interactive manner. It draws on many different bodies of knowledge and experience and is crucial in determining the success of the project. In addition to selecting an overall strategy and key team members to achieve the project's objectives, it determines the overall procurement approach and sets up the control systems that guide the project through to the final Post-completion Review and Project Close-out Report Stage. In particular, the Strategy Stage establishes the objectives for the control systems. These deal with much more than quality, time and cost. They provide agreed means of controlling value from the client's point of view, monitoring financial matters that influence the project's success, managing risk, making decisions, holding meetings, maintaining the project's information systems, and all the other control systems necessary for the project to be undertaken efficiently.

2nd

At the completion of the Strategy Stage, everything is in place for the Pre-construction Stage. This is when the design decisions are made. This stage includes statutory approvals and consents, and bringing manufacturers, contractors and their supply chains into the project team. Like the earlier stages, the Pre-construction Stage often requires many different professionals and specialists working in creative and highly interactive ways. It is therefore important that this stage is carefully managed using the control systems established during the Strategy Stage to provide everyone involved with relevant, timely and accurate feedback about their decisions. Completion of this stage provides all the information needed for construction to begin.

The Construction Stage is when the actual building or other facility that the client needs is produced. In modern practice this is a rapid and efficient assembly process delivering high-quality facilities. It makes considerable demands on the control systems, especially those concerned with time and quality. The complex nature of modern buildings and other facilities and their unique interaction with a specific site means that problems will arise and have to be resolved rapidly. Information systems are tested to the full, design changes have to be managed, construction and fitting out teams have to be brought into the team and empowered to work efficiently. Costs have to be controlled and disputes resolved without compromising the value and quality delivered to the client.

The Construction Stage leads seamlessly into a key stage in modern construction and development projects, [the Engineering Services Commissioning Stage.] The complexity and sophistication of modern engineering services make it essential that time is set aside to test and fine tune each system. Therefore these activities form a distinct and separate stage which should be finished before beginning the Completion, Handover and Occupation Stage which is when the client takes over the completed building or other facility.

The client's occupational commissioning needs to be managed as carefully as all the other stages because it can have a decisive influence on the project's overall success. New users always have much to learn about what a new building or other facility provides. They need training and help in making best use of their new building or other facility. It is good practice for their interests and concerns to be considered during the earlier stages and preparation for their move into the new facility begun early so there are no surprises when the client's organisation moves in.

The final stage is the Post-completion Review and Project Close-out Report Stage. This provides the opportunity for the project team to consider how well the project's objectives have been met and what lessons should be taken from the project. A formal report describing these matters provides a potentially important contribution to knowledge. For clients who have regular programmes of projects and for project teams that stay together over several projects, such reports provide directly relevant feedback. Even where this is not the case, everyone involved in a project team, including the client, is likely to learn from looking back at their joint performance in a careful objective review.

Part 1
Project management

LIVERPOOL JOHN MOORES UNIVERSITY
Aldham Robarts L.R.C.
TEL. 051 231 3701/3634

1 Inception stage

Introduction

Capital projects are usually complex, requiring significant management skills, co-ordination of a wide range of people with different expertise and ensuring completion within the parameters of time, value and necessary specifications.

The inception stage of any construction and development project requires the decision from the client that a potential project represents the best way of meeting a defined need.

In assessing the need for construction, key questions should include:

- Why is the project needed?
- How best can the need be fulfilled? (For example, a new building, or refurbishment, or extension of existing structure.)
- What benefits are expected as a result of the project?
- What are the investment/funding options?
- What risks related to the development can be foreseen at this stage?

Client's objectives

The main objective at this stage for the client is to make the decision to invest in a construction or development project. Clients should have a business case prepared (capital expenditure programme) involving careful analysis of their business, organisation, present facilities and future needs. Experienced clients may have the necessary expertise to prepare their business cases in-house. Less experienced clients may need help. Many project managers are able to contribute to this process. This process will result in a project-specific statement of need. The client's objective will be to obtain a totally functional facility, which satisfies this need and must not be confused with the project objectives, which will be developed later from the statement of need.

A sound business case prepared at this stage will:

- be driven by needs
- be based on sound information and reasonable estimation
- contain rational processes
- be aware of associated risks
- have flexibility
- maximise the scope of obtaining best value from resources
- utilise previous experiences.

Client's internal team

Investment decision maker: this is typically a corporate team of senior managers/ directors who review the potential project and monitor the progress. However, they are seldom directly involved in the project process.

Project sponsor: typically a senior person in the client's organisation, acting as the focal point for key decisions about progress and variations. The project sponsor has to possess the skills to lead and manage the client role, have the authority to make day-to-day decisions and have access to people who are making key decisions.

Client's advisor: the project sponsor can appoint an independent client advisor (also referred to as construction advisor or project advisor) who will provide professional advice in determining the necessity of construction and means of procurement, if necessary. If the client does not have the necessary skills in-house, an external consultant should be appointed. If advice is taken from a consultant or a contractor, those organisations have a vested interest not only in confirming the client's need, but also in selling their services and products.

The client advisor should understand the objectives and requirements of the client, but should not have a vested interest in any of the project options beyond the provision of expert advice to clients. He or she should not form part of any team and instead provide advice directly to the client. Other areas where the client may seek independent advice include: chartered accounting, tax and legal aspects, market research, town planning, chartered surveying and investment banking.

The client advisor can assist with:

- business case development
- investment appraisal
- understanding the need for a project
- deciding the type of project that meets the need
- generating and appraising options
- selecting an appropriate option
- risk assessment
- advising the client on the choice of procurement route
- selecting and appointing the project team
- measuring and monitoring performance.

Project manager

Project managers can come from a variety of backgrounds, but will need to have the necessary skills and competencies to manage all aspects of a project from inception to occupation. This role may be fulfilled by a member of the client's organisation or by an external appointment.

Project manager's objectives

The project manager, both acting on behalf of, and representing the client has the duty of 'providing a cost-effective and independent service, selecting, correlating, integrating and managing different disciplines and expertise, to satisfy the objectives and provisions of the project brief from inception to completion. The service provided must be to the client's satisfaction, safeguard his interests at all times, and, where possible, give consideration to the needs of the eventual user of the facility.'

The key role of the project manager is to motivate, manage, co-ordinate and maintain the morale of the whole project team. This leadership function is essentially about managing people and its importance cannot be overstated. A familiarity with all the other tools and techniques of project management will not compensate for shortcomings in this vital area.

In dealing with the project team, the project manager has an obligation to recognise and respect the professional codes of the other disciplines and, in particular, the responsibilities of all disciplines to society, the environment and each other.

There are differences in the levels of responsibility, authority and job title of the individual responsible for the project, and the terms project manager, project co-ordinator and project administrator are all used to reflect the variations.

It is essential that, in ensuring an effective and cost-conscious service, the project should be under the direction and control of a competent practitioner with a proven project management track record usually developed from a construction industry-related professional discipline. This person is designated the project manager and is to be appointed by the client with full responsibility for the project. Having delegated powers at inception, the project manager will exercise, in the closest association with the project team, an executive role throughout the project.

Project manager's duties

The duties of a project manager will vary depending on the client's expertise and requirements, the nature of the project, the timing of the appointment and similar factors. If the client is inexperienced in construction the project manager may be required to develop his or her own brief. Whatever the project manager's specific duties in relation to the various stages of a project are, there is the continuous duty of exercising control of project time, cost and performance. Such control is achieved through forward thinking and the provision of good information as the basis for decisions for both the project manager and the client. A matrix correlating suggested project management duties and client's requirements is shown in Table 1.1.

Typical terms of engagement for a project manager are given in Appendix 1. They will be subject to modifications to reflect the client's objectives, the nature of the project and contractual requirements.

A *supervising officer and/or contract administrator* may be appointed for the construction and subsequent stages of the project. This post is often taken by a member of the project team who will have a direct contractual responsibility to the client, subject to consultation with the project manager.

The term 'project co-ordinator' is applied where the responsibility and authority embrace only part of the project, e.g. pre-construction, construction and handover/migration stages. (For professional indemnity insurance purposes a distinction is made between project management and project co-ordination. If the project manager appoints other consultants the service is defined as project management. If the client appoints other consultants the service is defined as project co-ordination.)

Appointment of project manager

To ensure professional, competent management co-ordination, monitoring and controlling of the project right from the inception stage, and its satisfactory

Table 1.1 Suggested project manager's duties

Duties*	Client's requirements			
	In-house project management		Independent project management	
	Project management	Project co-ordination	Project management	Project co-ordination
Be party to the contract	●		+	
Assist in preparing the project brief	●		●	
Develop project manager's brief	●		●	
Advise on budget/funding arrangements	●		+	
Advise on site acquisition, grants and planning	●		+	
Arrange feasibility study and report	●	+	●	+
Develop project strategy	●	+	●	+
Prepare project handbook	●	+	●	+
Develop consultant's briefs	●	+	●	+
Devise project programme	●	+	●	+
Select project team members	●	+	+	+
Establish management structure	●	+	●	+
Co-ordinate design processes	●	+	●	+
Appoint consultants	●		●	+
Arrange insurance and warranties	●	●	●	+
Select procurement system	●	●	●	+
Arrange tender documentation	●	●	●	+
Organise contractor pre-qualification	●	●	●	+
Evaluate tenders	●	●	●	+
Participate in contractor selection	●	●	●	+
Participate in contractor appointment	●	●	●	+
Organise control systems	●	●	●	●
Monitor progress	●	●	●	●
Arrange meetings	●	●	●	●
Authorise payments	●	●	●	+
Organise communication/reporting systems	●	●	●	●
Provide total co-ordination	●	●	●	●
Issue safety/health procedures	●	●	●	●
Address environmental aspects	●	●	●	●
Co-ordinate statutory authorities	●	●	●	●
Monitor budget and variation orders	●	●	●	●
Develop final account	●	●	●	●
Arrange pre-commissioning/ commissioning	●	●	●	●
Organise handover/occupation	●	●	●	●
Advise on marketing/disposal	●	+	●	+
Organise maintenance manuals	●	●	●	+
Plan for maintenance period	●	●	●	+
Develop maintenance programme/ staff training	●	●	●	+
Plan facilities management	●	●	●	+
Arrange for feedback monitoring	●	●	●	+

*Duties vary by project, and relevant responsibility and authority.
Symbols: (●) = suggested duties; (+) = possible additional duties.

completion in accordance with the brief, it is advisable to appoint the project manager at a very early stage, possibly at the inception. However, depending on the nature and type of project and the client's in-house expertise, the project manager can be appointed at a later stage, at the feasibility or perhaps at the beginning of the strategy stage. In selecting and appointing the project manager, the client may follow the procedure for selecting and appointing consultants. Further information and advice on this subject is contained in *Successful Construction*, the code of practice for construction clients born out of the Latham review. The Construction Industry Council (CIC), ICE and the RICS have also published advice and guidance on this.

Managing people

Project management, although strongly associated with change management and systems, is above all about managing people. It is about motivating the project team, middle management and the workforce and gaining their commitment. It is also about achieving an effective form of relationship, which will enable an atmosphere of mutual co-operation to exist (see Sir Michael Latham's report *Constructing the Team* and Sir John Egan's report *Rethinking Construction*).

People: the most important resource

Although it is important to exploit new technology in order to achieve technological leadership and thus a competitive advantage, it is feasible that all firms could ultimately have access to similar technology. It is, therefore, the human resources that will make the difference and ultimately create the competitive advantage. Even computer-based systems are only as good as their designers and operators. People are our industry's most important resource.

It requires special skills to be successful at organising, motivating and negotiating with people. Although some people have a greater natural talent for this than others, everyone can improve their natural ability through appropriate education and training.

The skills the project manager will need to consider when assessing an individual may include the following.

■ What a person can do: skills, competencies.

■ What a person can achieve: output, performance.

■ How a person behaves: personality, attitudes, intellect.

■ What a person knows: knowledge, experience.

The skills the project manager will use during the course of a project will include:

■ Communication: using all means, the foremost skill.

■ Organising: using systems and good management techniques.

■ Planning: via accurate forecasting and scheduling.

■ Co-ordination: by liaising, harmonising and understanding.

■ Controlling: via monitoring and response techniques.

■ Leadership: by example.

■ Delegation: through trust.

■ Negotiation: by reason.

■ Motivation: through appropriate incentives.

■ Initiative: by performance.

■ Judgement: through experience and intellect.

Establishing objectives

The recognition that members of the project team have differing and sometimes conflicting objectives is the first step in ensuring that the team operates as an effective unit.

With the client's project objectives in sharp focus, attention is directed towards overcoming any conflict in the aims of team members. Presentation of objectives, team selection, choice of working environment, definition of levels of responsibility, authority and communication procedures; all are influential in ensuring that team members meet their personal objectives as part of the successful execution of the project.

The project manager should aim to create an environment in which the client and all his or her team members can achieve their personal, as well as project, goals.

There is no doubt that team performance is optimised when members are encouraged to identify and tackle problems early in the process. This will only occur when the benefits of revealing mistakes and omissions outweigh any penalties imposed by the client. Promotion of an open, 'blame-free' culture, where the project manager leads by example, will also help in breaking down communication barriers.

2

Feasibility stage

Client's objectives

The objectives for the client at this stage include specifying project objectives, out-lining possible options and selecting the most suitable option through value and risk assessment. Establishing the project execution plan for the selected option should be the key output at this stage.

Outline project brief

For most clients a construction project is necessary to satisfy their business objectives. The client's objectives may be as complex as the introduction and accommodation of some new technology into a manufacturing facility or the creation of a new corporate headquarters; or they may be as simple as obtaining the optimum return on resources available for investment in a speculative office building.

The client's objectives are usually formulated by the organisation's board or policy-making body (the investment decision maker) and may include certain constraints – usually related to time, cost, performance and location. The client's objectives must cover the function and quality of the building or other facility.

If it is considered that the objectives are of a complexity or size to merit the engage-ment of a project manager, the appointment should ideally be made as early as possible, preferably after approving the project requirements at the inception stage. This will ensure the benefit of the special expertise of the project manager in help-ing to define the objectives and in devising and assessing options for the achieve-ment of the objectives.

The project manager should be provided with or assist in preparing a clear statement of the client's objectives and any known constraints. This is the initial outline project brief to which the project manager will then work.

A typical example of a template for an outline project brief is shown in Figure 2.1.

There is seldom, if ever, a single route available for the achievement of the client's objectives, so the project manager's task is to work under the client's direction to help establish a route which will best meet the client's objectives within the constraints that are set. In liaison with the client the project manager will discuss the available options and initiate feasibility studies to determine the one to be adopt-ed. In order that the feasibility studies are effective, the information used should be as full and accurate as possible.

Much of that information will need to be provided by specialists and experts. Some of these experts may be available within the client's own organisation or be regu-larly retained by the client – lawyers, financial advisers, insurance consultants and the like. Others, such as architects, engineers, planning supervisors, town planning

PROJECT TITLE

PROJECT REF

CUSTOMER: *(internal/external)*

PROJECT SPONSOR:

PROJECT MANAGER:

GOAL

THIS NEEDS TO BE SPECIFIC AND INCLUDE THE JUSTIFICATION FOR THE PROJECT

It should spell out: **what** *will be done and by* **when***;*

OBJECTIVES

It is essential these cover the OUTCOMES expected of the project and that preferably they are:

Specific – i.e. clear and relevant

Measurable – i.e. so it is feasible to see when it is happening

Achievable / agreed to – helpful to use positive language and that others 'buy-in' to the objectives

Realistic – this depends on three factors: resources / time / outcome or aim

Time bound – have a time limit – without this they are wishes

APPROACH

The project plan should include the key milestones for the review, i.e. set a target date for agreeing the project brief and target dates for completing key stages of the project.

SCOPE

THIS SETS THE PROJECT BOUNDARIES AND IT CAN BE USEFUL TO ADD WHAT IS NOT COVERED.

It can be a useful reference point if the project changes in due course

CONSTRAINTS

Could add 'start' date and 'end' dates here

It is particularly important in the context of best value, to identify here genuine constraints rather than customer preconceived ideas about the solution

DEPENDENCIES

This identifies factors outside the control of the project manager, and may include:

- Supply of information
- Decisions being taken at the right time
- Other supporting projects

RESOURCE REQUIREMENTS

Include estimate of project days and costs

AGREED

Signature: Date:

Project Manager:

Project Sponsor :

Note: The above example of the possible template of an outline project brief is for guidance purposes only.

Figure 2.1 Outline project brief template

consultants, land surveyors and geotechnical engineers, may need to be specially commissioned.

Feasibility study reports should include:

- Scope of investigation (from outline project brief) including establishing service objectives and financial objectives.

- Studies on requirements and risks.

- Public consultation (if applicable).

- Geo-technical study (if applicable).

- Environmental impact assessment.

- Health and safety study.

- Legal/statutory/planning requirements or constraints.

- Estimates of capital and operating costs (demolition costs, if applicable).

- Assessment of potential funding.

- Potential site assessments (if applicable).

The client will commission feasibility studies and establish that the project is both deliverable and financially viable. The client should have already instructed the project manager at this stage and if so, his or her input will be made alongside the reports and views of the various consultants.

The client may ask the project manager to engage and brief the various specialists for the feasibility studies, co-ordinate the information, assess the various options and report conclusions and recommendations to the client. The feasibility report should include a 'risk assessment' for each option and will usually also determine the contractual procurement route to be adopted and a draft master schedule applicable to each. The client may also require comparative 'life-cycle costings' to be included for each option.

It is at this stage that the end value or outputs of the development must be assessed. Accurate and well-informed assessments of revenue streams and prospective capital values must be made with the expert help of specialist consultants and valuers. If the proposed project does not pass these tests, then changes will have to be made. At this point the project manager plays a crucial and pivotal role in advising the non-cognate client that proper attention should be paid to the specialist advice provided on the value side of the cost/value equation. Where organisations have their own in-house assessment team, it is assumed that they can take their own decisions on the financial feasibility of a project.

During the progress of the feasibility studies the project manager will convene and minute meetings of the feasibility team, report progress to the client and advise the client if the agreed budget is likely to be exceeded. Feasibility studies are the most crucial, but also the least certain, phase of a project. Time and money expended at this stage will be repaid in the overall success of the project. The specialists engaged for the feasibility studies are most commonly reimbursed on a time–charge basis and without commitment to engage the specialist beyond the completion of the feasibility study, although often some or all members of the feasibility team will be invited to participate in the selection process to become design team members.

The project manager will obtain the client's decision on which option to adopt for the project, and this option is designated the outline project brief. The process of developing the project brief from the client's objectives is shown in Figure 2.2.

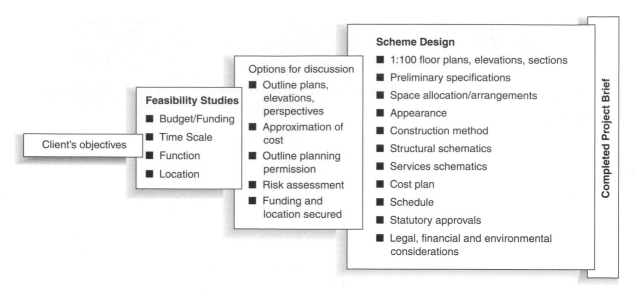

Figure 2.2 Development of project brief from objectives

Site selection and acquisition

Site selection and acquisition is an important stage in the project cycle in the situation where the client does not own the site to be developed. It should be effected as early as possible and, ideally, in parallel with the feasibility study. The work is carried out by a specialist consultant and monitored by the project manager.

The objectives are to ensure that the requirements for the site are defined in terms of the facility to be constructed, that the selected site meets these requirements and that it is acquired within the constraints of the project schedule and with minimal risk to the client.

To achieve these objectives the following tasks need to be carried out:

■ Preparing a statement of objectives/requirements for the site and facility/ buildings and agreeing this with the client.

■ Preparing a specification for site selection and criteria for evaluating sites based on the objectives/requirements.

■ Establishing the outline funding arrangements.

■ Determining responsibilities within the project team (client/project manager/ commercial estate agent).

■ Appointing/briefing members of the team and developing a schedule for site selection and acquisition; monitoring and controlling progress against the schedule.

■ Actioning site searches and collecting data on sites, including local planning requirements, for evaluation against established criteria.

■ Evaluating sites against criteria and producing a short list of three or four; agreeing weightings with the client.

■ Establishing initial outline designs and developing costs.

■ Discussing short-listed sites with relevant planning authorities.

■ Obtaining advice on approximate open-market value of short-listed sites.

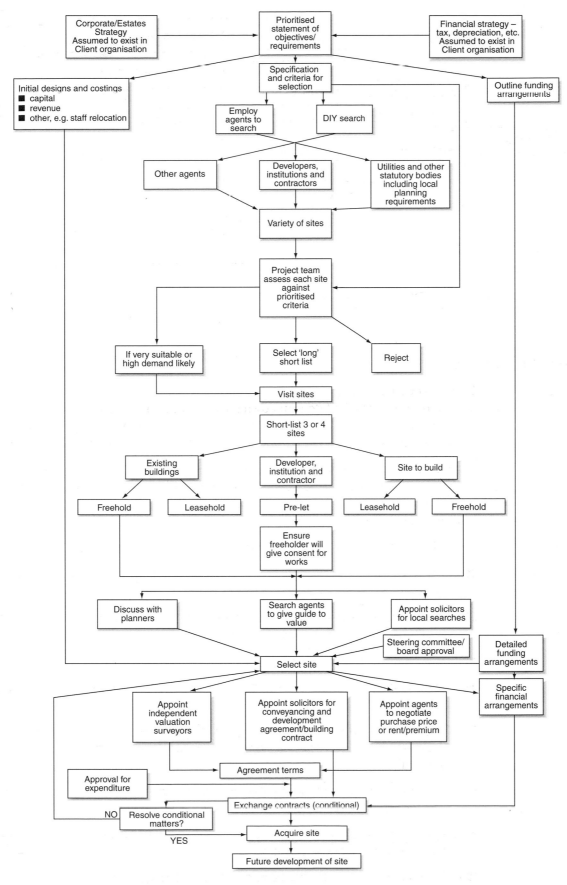

Figure 2.3 Site selection and acquisition

- Selecting the site from a short list.

- Appointing agents for price negotiation and separate agents for independent valuation.

- Appointing solicitors as appropriate.

- Determining specific financial arrangements.

- Exchanging contracts for site acquisition once terms are agreed, conditional upon relevant matters, e.g. ground investigation, planning consent.

Detailed project brief

The formulation of the detailed brief for the project is an interactive process involving most members of the design team and appropriate representatives of the client organisation. It is for the project manager to manage the process, resolving conflicts, obtaining client's decisions, recording the brief and obtaining the client's approval.

Table 2.1 Suggested contents for detailed project brief

The following is a suggested list of contents, which should be tailored to the requirements and environment of each project.

- Background
- Project definition, explaining what the project needs to achieve. It will contain:
  1. Project objectives
  2. Project scope
  3. Outline project deliverables and/or desired outcomes
  4. Any exclusions
  5. Constraints
  6. Interfaces
- Outline business case
  1. A description of how this project supports business strategy, plans or schedules
  2. The reason for selection of this solution
- Customer's quality expectations
- Acceptance criteria
- Risk assessment

If earlier work has been done, the project brief may refer to the document(s) containing useful information, such as the outline project brief, rather than include copies of them.

It is not unusual during this phase for the client to modify his thinking on various aspects of the proposals, and there is certainly the opportunity and scope for change during this phase. Figure 2.4 demonstrates graphically the relationship between 'scope for change' and the 'cost of change' set against the time-scale of a development. It will be seen that the crossover point occurs at the completion of the strategy stage. The client's attention should always be drawn to this relationship and to the benefits of brief and design freezes.

The key emphasis for the client should be to understand and establish enough information about the end requirements and objectives for developing the project. This point cannot be overemphasised. It is essential for the project manager to

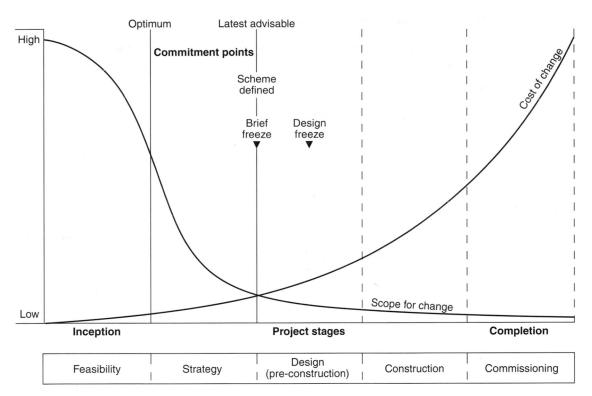

Figure 2.4 Relationship between scope for change and cost of change

identify the client's needs and objectives through careful and tactful examination, in order to minimise potential future changes to the project brief. Many clients who are unfamiliar with the development process are not perhaps fully aware of the benefit of getting the design right as far as possible at the start of the project, and therefore the importance of making sure the brief fully reflects the client's requirements, before materials are ordered and construction commences. The project manager should do his utmost, therefore, to familiarise the client with the potential cost and time implications of design changes and identify as clearly as possible the precise requirements of the client. The project manager may give this advice but the Latham Report and the Strategic Construction Forum (SCF) advocate the appointment of a 'client adviser' to give independent advice until the need for a construction project has been established.

Detailed design brief

Within the detailed project brief, the assembly of the detailed design brief will normally be the responsibility of the lead design consultant along with the project manager and, where appropriate, the client.

The project manager will monitor the assembly of the detailed design brief to ensure compliance with the outline project brief, the project budget and the master schedule.

Depending on the procurement method adopted and the master schedule, the assembly of the detailed design brief may occur in parallel with other activities, such as the development of the final scheme design and site preparation. A decision on some elements of the detailed design brief may be deferred by the client even until after construction has commenced. This is risky and should be allowed only when time is of over-riding concern to the client. It means that the design of the first elements to be constructed have to be over-designed to allow for any possible subsequent client requirements. It is much better and usually possible (except in extreme emergencies) to complete the brief and the design before construction begins. It just needs good project management.

The project manager will advise the client of the implications for cost, time and risk in the deferment of any elements of the detailed design brief.

The project manager will monitor the progress of the assembly of the detailed design brief and notify the client of the effects on cost, time, quality, function and financial viability of any changes from the outline design brief. The detailed design brief, or such part of it that has not been deferred, having been tested against these criteria, should be presented to the client as a formal document for his approval. The lead design consultant and the project manager should normally make the presentation jointly.

Scheme design

Once approved by the client, the detailed design brief becomes the control document for the design and will be issued by the project manager to all members of the design team. The project manager, with the approval of the client, will instruct the design team to complete the scheme design.

In order to make informed decisions on deliverability of a project and also its financial viability, it will be necessary to instruct the architect (the cost consultant and other experts as appropriate) to prepare site layouts, floor plans, elevations and other drawings in sufficient detail for the cost consultant to prepare preliminary cost plans.

The project manager will monitor the completion of the scheme design, arrange for cost checks benchmarking against the cost plan (see Chapter 3 for details), and obtain confirmation that the design meets the detailed design brief and all external constraints.

Funding and investment appraisal

In all development projects, a balance between cost and value must be established. The financial appraisal of the project can be assessed either by calculating the total cost and then assessing the value or alternatively, calculating the value of the end product and working out the project costs with an eye to value. In either case, the client will expect value to exceed cost and in the case of developer-led projects, the client will, at inception stage, have decided on the level of profit they will require for the amount of risk involved. A thorough risk analysis, particularly analysing the market conditions on the potential revenue generation, interest rate changes, potential impact of schedule delay and outcomes of similar historical precedents (which may be incorporated within the business plan or development appraisal for the project), is usually performed to assist in decision making. Developers and many clients experienced in construction procurement may not require specific help from the project manager in these areas but should keep him or her well informed of the financial arrangements so that they can be taken into account in any project decisions. On the other hand, clients unfamiliar with construction may require an input by the project manager or from an independent construction advisor. In any case, although the project manager may have knowledge of project finance, it is unlikely that he or she will be expected to advise in this area. Specialist advisors or the client will arrange bank finance; tax and legal advice should be sought in all those areas relating to the acquisition of the site and the financing of the development project. The project manager should be able to advise on certain matters relating to VAT, budgetary systems, cost and cash flow. Project managers should also know when and where to go for specialist advice to augment their own expertise or their client's expertise in such matters.

Project/market suitability

The key to a successful project is to try to bring together all the various elements into a workable and viable whole. For validating projects in the commercial occupational market, good market awareness and the ability to judge not only occupiers' requirements but also trends in the investment market are absolutely vital and a key issue at this stage is to ensure site selection, appropriate to those demands.

This test should apply whether the project is a shopping centre, hotel, office block, hospital, library, stadium or business and industrial park. In buildings for overtly commercial use, relatively simple tests can be made to assess the project's suitability for the target market and thence the project's overall viability. In other cases such as buildings for public or leisure uses, different and more complex tests must be applied to ensure that the revenues and outputs offer a satisfactory return on costs. Independent advice on these revenue/value issues is important and the client should look to the project manager to point him or her in the right direction. This will enable a valid assessment of the risk analysis.

Project suitability also encompasses impact analysis on a company's financial performance, whether it be impact on balance sheet, profit and loss or cash flow, and an informed assessment of these matters needs to be carried out by the in-house team with proper support from the project manager.

At this stage, the project manager will also be ensuring that the brief ties in with the business assumption of the client and where it does not, pointing the client in the right direction.

Decision to go ahead

The client, after reviewing the documents generated throughout this phase, has to reaffirm the decision to proceed with the project, in order to:

- Provide the authorisation of financial management and control throughout the project.

- Ensure that no commitment is made to large expenditure on the project before verifying that it has been authorised as required.

Table 2.2 Client's decision prompt list

■ Is there adequate funding for the project?
■ Does the detailed project brief demonstrate the existence of a worthwhile project, and hence justify the investment involved?
■ Are external support and facility requirements available and committed?
■ Have the most appropriate standards been applied, in order to achieve the best value for money?
■ Are assurance responsibilities allocated and accepted?

At the early stages of a project it is unlikely that the actual costs will be known. It is important to check that the need for financial provision has been recognised by all the parties. The actual project cost, which perhaps will almost certainly be higher than the original estimate, requires the question of affordability to be revisited at that stage to be sure that adequate funds will be available.

Project execution plan (PEP)

The PEP is the core document for the management of a project. It is a statement of policies and procedures defined by the project sponsor, although usually developed by the project manager for the project sponsor's approval. It sets out in a structured format the project scope, objectives and relative priorities.

This is a live document that enforces discipline and planning with a wider circulation than the project design team. It forms a basis for:

■ Sign off by the client body at the end of the feasibility and strategy phases.

■ A prospectus for funding.

■ An information and 'catch up' document for prospective contractors.

Some of the confidential information in the client version will be taken out of the published version to other parties.

Checklist for the PEP

Does the PEP:

■ include plans, procedures and control processes for project implementation and for monitoring and reporting progress?

■ define the role and responsibilities of all project participants, and is it a means of ensuring that everyone understands, accepts and carries out their responsibilities?

■ set out the mechanisms for audit, review and feedback, by defining the reporting and meetings requirements, and, where appropriate, the criteria for independent external review?

Essential contents

Much of a PEP will be standardised, but the standard will need to be modified to meet the particular circumstances of each project. A typical PEP might cover the items listed below, although some may appear under a number of headings with a cross-reference system employed to avoid duplication:

■ Project definition and brief.

■ Statement of objective.

■ The business plan with costs, revenues and cash flow projections including borrowings interest and tax calculations.

■ Market predictions and assumptions in respect of revenue and returns.

■ Functional and aesthetic brief.

■ Client management and limits of authority including the project manager.

- Financial procedures and delegated authority to place orders.
- Development strategy and procurement route.
- Risk assessment.
- Schedule and phasing.
- The scope content of each consultant appointment.
- Reconciled scheme design and budget.
- Detailed design.
- Package design and tendering.
- Construction.
- Commissioning and handover.
- Operation.
- Safety and environmental issues, such as the construction design and management regulations.
- Quality assurance.
- Post-project evaluation.

The PEP will change as a project progresses through its design and construction stages. It should be a dynamic document regularly updated and referred to as a communication tool, as well as a control reference.

3 Strategy stage

Client's objectives

The main aims for the client at this stage include setting up the project organisation, establishing the procurement strategy and commissioning/occupation issues through identifying project targets, assessing and managing risks and establishing the project plan.

Interlinking with feasibility

Distinction between the tasks and activities of the feasibility and strategy stages is not always clear, as each is influenced to a certain extent by the considerations and findings of the other. The two tasks and activities need to relate to each other in order to achieve effective outcomes for both. Feedback (Figure 3.1) is essential in order to establish for the client a sound basis for decision making at the conceptual phases of the project and, subsequently, for its effective execution. The order in which the activities are set out here is not significant and will vary for specific projects.

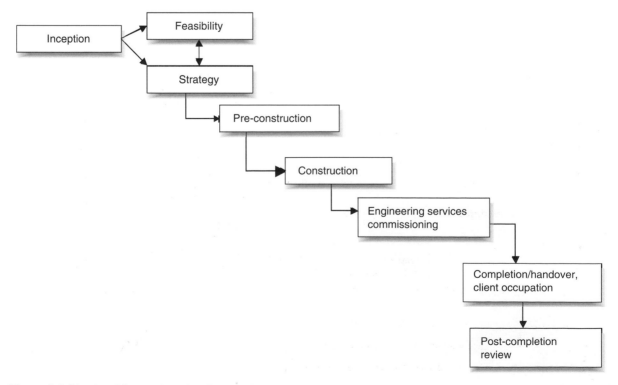

Figure 3.1 Stages of the project development

Project team structure

Projects are usually carried out by a project team under the overall direction and supervision of a project manager. The team normally comprises:

■ Client's internal team (appropriate representatives).

■ Project manager (either within the client's own organisation or independently appointed).

■ Design team: architects, structural/civil/mechanical and electrical (M&E) engineers and technology specialists.

■ Consultants covering quantity surveying, development surveying, planning, legal issues, valuation, finance/leasing, insurances, design audit, health and safety and environmental protection, access issues, facilities management, highways/traffic planning, construction management and other specialisms.

■ Contractors and subcontractors.

The project team structure for project management is shown in Figure 3.2. This structure is idealised and in practice there will be many variants, depending on the

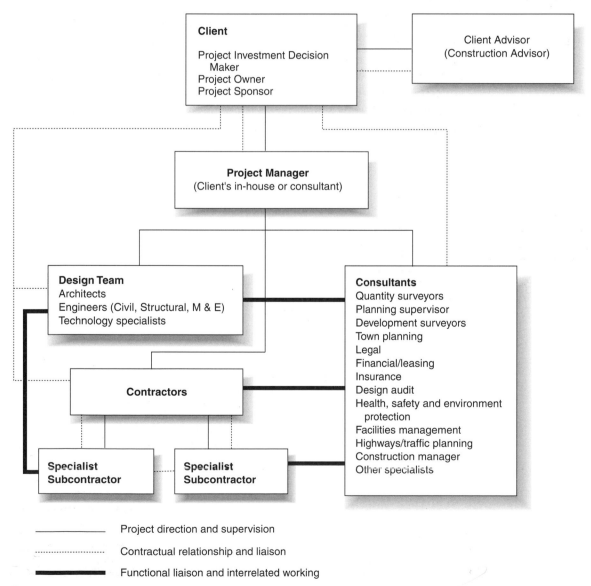

Figure 3.2 Project team structure

nature of the project, the contractual arrangements, type of project management (external or in-house) involved, and above all, the client's requirements. It should be one of the duties of the project manager to advise the client on the most appropriate project team structure for a particular project.

Effective project management must, at all times, fully embrace all provisions for quality assurance, financial control, health and safety, access provision and environmental protection. These aspects are to be considered as incorporated and implied in all relevant activities specified in this code.

Selecting the project team

When establishing a project team, different skills will have to be considered. During selection the project manager has to attain the following factors:

- A commitment by the project team to clearly defined and measurable project objectives.

- Firm duties of teamwork, with shared financial motivation to pursue those objectives. These should involve a general presumption to achieve 'win–win' solutions to problems which may arise during the course of the project. (Further guidance towards a co-operative and mutually beneficial team working arrangement is available in various industry publications, notably those published by the CBPP, CIC and CIRIA – see the bibliography. Almost all the professional institutions including RICS, RIBA, ICE and CIOB have also published guidance and information documents in this area. Also see the Latham Report, the Egan Report, the OGC guidance and the NAO report on construction.)

- The production of satisfactory evidence from each team member, to show that they can contribute effectively to the project objectives. This evidence may include a realistic schedule, a financial plan and a demonstration of adequate resources.

- When choosing each team member, as suggested in Chapter 1, special attention is to be paid to their:
  - relevant experience
  - technical qualifications
  - appreciation of project objectives
  - level of available supporting resources
  - creative/innovative ability
  - enthusiasm and commitment
  - positive team attitude
  - communication skills.

- Financial strength and core resource strength can also be important. In choosing contractors it is wise to avoid placing orders with companies if the order is greater than 20% of annual turnover.

- Defining clear lines of communication between the respective project team members.

- Promoting a working environment which encourages an interchange of ideas by rewarding initiatives which ultimately benefit the project.

- Undertaking regular performance appraisals for all project team members.

- Ensuring that project team members are suitably located and that communication protocols have been established (particularly for electronic sharing of information) so as to facilitate regular contact with each other, as well as with their own organisations.

- Defining clear areas of responsibility and lines of authority for each project team member, and communicating these within the team.

- Identifying a suitable deputy for each team member, who will be sufficiently familiar with the project to be able to act as their replacement if necessary.

- Making provision for project team members to meet informally and socially outside the work environment, on a regular basis, in team building activities.

Strategy outline and development

A typical strategy stage consists of the main elements shown in Figure 3.3.

The project manager performs several principal activities at this stage which may include all, or most of the following:

- Reviewing, and in some cases developing, the detail project brief with the client and any existing members of the project team to ascertain that the client's objectives will be met. Preparing a final version in written form with supplementary appendices where these add to the general understanding of the issues that support the brief itself.

- Establishing, in consultation with the client and other consultants, a project management structure (organisation) and the participants' roles and responsibilities, including access to client and related communication routes, and

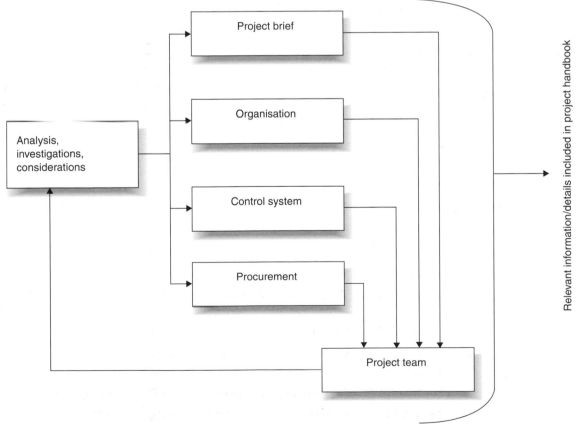

Figure 3.3 Elements of the strategy stage

'decision required' points (see Part 2 for details). This should be developed and presented in the project documents for the reference of all parties.

■ Ensuring, in liaison with the client, the planning supervisor, design consultants and the principal contractor when appointed, that appropriate arrangements have been made to meet the requirements of the Construction (Design and Management) Regulations. Key duties under the regulations are summarised in Appendix 2, Table 1. Other guidance about relevant appointments and documentation is listed in Appendix 2, Tables 2 and 3, respectively.

■ Establishing that 'value management' is applied effectively from the earliest stages of the preparation of the design brief until the design is complete. The emphasis should be on providing value for money and in producing a building/facility that can be constructed and operated at the lowest cost without reducing quality, scope or specification. The design team and consultants should be encouraged not to accept conventional wisdom on what buildings/facilities cost, but consciously seek to reduce cost by better design and construction methods. An approach where the whole team 'designs in quality and drives out cost' at all stages in the design process should be encouraged – emphasis on overall value should be encouraged. Further guidance on value management is included Chapter 4 and in Appendix 10.

■ Advising the client on the recruitment and appointment of additional consultants and design team members, i.e.

   ○ preparation of appropriate definition of roles and responsibilities

   ○ preparation and issue of selection/tender documentation

   ○ evaluation, reporting and making recommendations

   ○ assisting the client in the preparation of agreements and in selection and appointment.

■ Drawing the client's attention to the benefits of project insurance for the whole project and works (currently obtainable at 0.5% per £1m value of construction), as suggested by the Egan Report *Accelerating Change* and providing assistance in assessing the risks on the project and including an appropriate contingency sum in the project budget.

■ Putting in place procedures for managing risk as a continuous project activity. A project risk assessment checklist (Appendix 9) may be used or adapted as part of such a procedure. (These risks are not to be confused with risks covered by the CDM Regulations, although CDM risks will form a subset of an overall risk management regime.)

■ Selection, or development, and agreement of the most appropriate form of contract relative to the project objectives and the parameters of cost, time, quality, function and financial viability.

■ Assisting the client in completing site selection/evaluation, investigation and acquisition.

■ Advising on whether certain activities, such as fitting out and occupation/migration, constitute separate projects and should be treated as such.

■ Making the client aware of relevant statutory submissions and other consultations that may be required in the delivery of the project.

Project organisation and control

A project management organisation structure sets out unambiguously and in detail how the parties to the project are to perform their functions in relation to each other in contributing to the overall scheme. This should be recorded in the project handbook. It also identifies arrangements and procedures for monitoring and controlling the relevant administrative details. It is updated as circumstances dictate during the lifetime of the project, and should allow project objectives and success criteria to be communicated and agreed by all concerned and promote effective teamwork.

Procedures covering the relationships and arrangements for monitoring, control and administration of the project should be developed, with the assistance of parties involved, for all stages of the project and cover time, costs, quality and reporting/decision-making arrangements.

The organisation structure should clearly identify the involvement and obligations of the client and their organisational backup.

Information technology

It is usual for extensive use to be made of computer applications as tools to assist most project management functions. It is essential for project managers to keep abreast of developments in this area in order to select and recommend appropriate packages and communication protocols for use on a project. It is particularly important to make sure that systems used by project team members are compatible to facilitate electronic exchange of data. E-mail, project specific websites, integrated project data applications and teleconferencing are examples of tools which may be required, and the project manager will need to be able to define the ways in which these tools are used and how the transfer of information is managed and monitored. Examples and further guidance towards successfully utilising information and communication technology in construction projects are available from ITCBP publications. (For guidance regarding project management software see Appendix 12.)

Project planning

The project master schedule should be developed and agreed with the client and the consultants concerned, and detailed schedules for each stage of the project should be prepared as soon as the necessary parameters are established (see Appendix 3).

While preparing the master schedule necessary allowances should be incorporated to provide for potential delays (including possible impact on initial revenue generation) in activities such as applying and obtaining statutory approvals, external consultations and enquiries, legal and funding negotiations and any other third-party agreements.

It is the project manager's responsibility to monitor the progress of the project against the master and stage schedule, identify risks to progress and to initiate necessary action to rectify potential or actual non-compliance.

Cost planning

A development budget study is undertaken to determine the total costs and returns expected from the project. A cost plan is prepared to include all construction costs, and all other items of project cost including professional fees and contingency. All costs included in the cost plan will also be included in the development budget in addition to the developer's returns and other extraneous items such as project insurance, surveys and his or her agent's or other specialist advisers' fees.

The objective of the cost plan is to allocate the budget to the main elements of the project to provide a basis for cost control. The terms *budget* and *cost plan* are often regarded as synonymous. However, the difference is that the *budget* is the limit of expenditure defined for the project, whereas the *cost plan* is the definition of what the money will be spent on and when. The cost plan should, therefore, include the best possible estimate of the cash flow for the project and should also set targets for the future running costs of the facility. The cost plan should cover all stages of the project and will be the essential reference against which the project costs are managed.

The method used to determine the budget will vary at different stages of the project, although the degree of certainty should increase as more project elements become better defined. The budget should be based on the client's business case and should change only if the business case changes. The aim of cost control is to produce the best possible building within the budget.

The cost plan provides the basis for a cash flow plan, based upon the master schedule, allocating expenditure and income to each period of the client's financial year. The expenditures should be given at a stated base-date level and at out-turn levels based on a stated forecast of inflation. A cash flow histogram and cumulative expenditure graph are shown in Figure 3.4.

Operational cost targets should be established for the various categories of running costs associated with the facility. This should accompany the capital cost plan and be included in the brief to consultants. Revenue, grants and tax planning for capital allowances must also be taken into consideration.

When the cost plan is in place it serves as the reference point for the monitoring and control of costs throughout a project. The list which follows should be used as an aid in setting up detailed cost control procedures for all stages of a project.

Cost control

The objective of cost control is to manage the delivery of the project within the approved budget. Regular cost reporting will facilitate, at all times, the best possible estimate of:

■ established project cost to date

■ anticipated final cost of the project

■ future cash flow.

In addition cost reporting may include assessments of:

■ ongoing risks to costs

■ costs in the use of the completed facility

■ potential savings.

Monitoring of expenditure to any particular date does not exert any control over future expenditure and, therefore, the final cost of the project. Effective cost control is effectively achieved when the whole of the project team has the correct attitude to cost control, i.e. one which will enable fulfilment of the client's objectives.

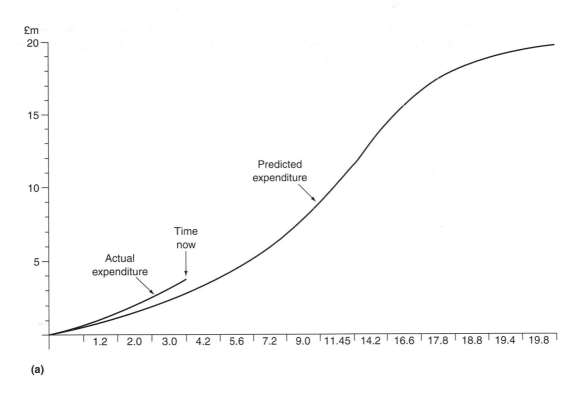

(a)

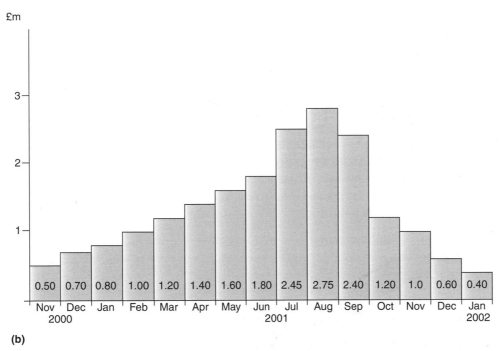

(b)

Figure 3.4 Examples of: (a) construction expenditure graph; (b) cash flow histogram

Effective cost control will require the following actions to be taken:

■ Establishing that all decisions taken during design and construction are based on a forecast of the cost implications of the alternatives being considered, and that no decisions are taken whose cost implications would cause the total budget to be exceeded.

■ Encouraging the project team to design within the cost plan, at all stages, and adopt the variation/change and design development control procedures for the project. It is generally acknowledged that 80% of cost is determined by design and 20% by construction. It is important that the project team is aware that no member of the team has the authority to increase costs on its section or element of the work. Increased costs on one item must always be balanced by savings on another.

- ■ Regularly updating and reissuing the cost plan and variation orders causing any alterations to the brief.

- ■ Adjusting the cash flow plan resulting from alterations in the target cost, the master schedule or the forecast of inflation.

- ■ Developing the cost plan in liaison with the project team as design and construction progress. At all times it should comprise the best possible estimate of the final cost of the project and of the future cash flow. Adherence to design freezes will aid cost control. (Development also means adding detail as more information about the work is assembled, replacing cost forecasts with more accurate ones or actual costs whenever better information can be obtained.)

- ■ As part of risk management, reviewing contingency and risk allowances at intervals and reporting the assessments are essential. Development of the cost plan should not involve increasing the total cost.

- ■ Checking that the agreed change management process is strictly followed at all stages of the project is very important (see Appendix 13). The procedure should only be carried out retrospectively and then only during the construction phase of the project, when it can be shown that otherwise significant delay, cost or danger would have been incurred by awaiting responses.

- ■ Arranging that the contractor is given the correct information at the correct time in order to minimise claims. Any anticipated or expected claims should be reported to the client and included in the regular cost reports.

- ■ Contingency money based on a thorough evaluation of the risks is available to pay for events which are unforeseen and unforeseeable. It should not be used to cover changes in the specification or in the client's requirements or for variations resulting from errors or omissions. Should the consultants consider that there is no alternative but to exceed the budget, a written request to the client must be submitted and correct authorisation received. This must include the following:

  - ○ details of variations leading to the request

  - ○ confirmation that the variations are essential

  - ○ confirmation that compensating savings are not possible without having an unacceptable effect on the quality or function of the completed project

- ■ Submitting regular, up-to-date and accurate cost reports to keep the client well informed of the current budgetary and cost situation.

- ■ Establishing that all parties are clear about the meaning of each entry in the cost report. No data should be incorrectly entered into the budget report or any incorrect deductions made from it.

- ■ Ensuring that the project costs are always reported back against the original approved budget. Any subsequent variations to the budget must be clearly indicated in the cost reports.

- ■ Plotting actual expenditure against predicted to give an indication of the project's progress (see Figure 3.4).

Procurement

In the context of this code of practice, procurement should be considered to be the process of identification, selection and commissioning of the contributions required for the construction phase of the project. The alternative methods of procurement referred to reflect the different organisational and contractual arrangements which can be made to ensure that the appropriate contributions are properly commissioned and that the interests of the client are safeguarded.

The various procurement options available reflect fundamental differences in the allocation of risk and responsibility to match the characteristics of different projects, therefore selection of the procurement option must be given strategic consideration. The project manager should advise on the relative benefits and disadvantages of each option, related to the particular circumstances of the project, for the benefit of the client.

The final choice of procurement method should be made on the basis of the characteristics of the project, the client and his requirements. The selection of method should be made when consideration is being given to the appointment of design and other specialist consultants because each option can have a different impact on the terms of appointment of the members of the project team.

The various procurement methods which may be pursued can be broadly classified under four headings:

o   traditional

o   design and build

o   management contracting

o   construction management.

Each method has its own variations. No method is best in all circumstances. They bring different degrees of certainty and risk to the project construction and development.

Traditional
The contractor builds to a defined scope of work for a fixed price lump sum regardless of costs. The client, however, remains responsible for the design and the performance of consultants under the building contract. The client appoints a design team, including a quantity surveyor responsible for financial and contractual advice. A building contractor is appointed, usually after a tender process, and usually based on one of the standard forms of contract, to carry out the construction. The tender process can be based on complete design information or partial design information plus provisional guidance if an early construction start is required.

Design and build
The client appoints a building contractor, usually on a standard form of contract to provide the completed building to agreed cost and schedule. The contractor is responsible for design and construction as defined in formal documentation known as the client's requirements. The appointment may be made after a tendering process incorporating variations on the method, or through negotiation. The client may appoint a consultant to oversee matters on his behalf. This arrangement transfers maximum risk to the contractor and generally has a good reputation for controlling schedule and the client's cost. The design, however, will be the most commercial response that a contractor can produce to satisfy the contract conditions.

The design and build contractor may be appointed when part of the design has been completed, and in these circumstances, the appointments of the design team may be formally passed on (contractually novated) to the design and build contractor. However, research has shown that this practice usually leads to potential conflict and poor quality and thus is not advisable.

LIVERPOOL JOHN MOORES UNIVERSITY
LEARNING & INFORMATION SERVICES

Prime contracting is an extension of the design and build concept. The prime contractor will be expected to have a well-established relationship with a supply chain of reliable suppliers. The prime contractor co-ordinates and manages throughout the design and construction period to provide a facility, which is fit for the specified purpose, and meets its predicted through-life costs. The prime contractor is paid all actual costs plus profit incurred in respect of measured work and design fees; it is only at risk in respect of its staff and preliminaries.

Public private partnerships (PPPs), particularly **private finance initiative** (PFI) projects, are developed for the provision of services that are required as a result of client needs and requirements and not specifically for the exclusive provision of capital assets such as buildings. For this reason it is preferable to investigate PPPs as soon as possible after a user need has been identified rather than leaving it until a conventional construction project has been selected as the solution. It is possible that a PPP may result in a solution (provision of services to meet the user need or objectives) that does not require a construction project at all.

One major benefit of this type of procurement is that the risks associated with providing the service are transferred to those best able to manage them. To achieve the project objective, the outputs that the service is intended to deliver (as a result of the facility/development) must be clearly defined at the initial stages by the client.

Framework agreements with a single supplier or a limited number of suppliers can result in significant savings to both parties. These agreements may cover prime contracting and design and build procurement routes. However, they are unlikely to be appropriate for clients that only occasionally have projects. They can be particularly appropriate for facilities management and maintenance requirements.

The expectation is that savings will come from: the absence of a requirement for reprocuring for each individual project, continuous improvement by transferring the learning from one project to another, reduced confrontation through extended co-working, and continuous workflow by keeping the same project team.

Management contracting

The client appoints a design team with responsibilities as in the traditional method and augmented by a management contractor whose expertise and advice is available throughout the design development and procurement processes. Specialist works subcontractors, who are contracted to the management contractor on terms approved by the contract administrator who may be the architect, the quantity surveyor or the project manager, carry out the construction. The appointments of the management contractor and the trade subcontractors are usually made on standard contract forms. The management contractor is reimbursed all their own costs and paid a percentage on project costs in the form of a guaranteed profit or fee.

Construction management

Construction management requires that the specialist works contractors are contracted to the client directly, involving the construction manager as a member of the project team acting as an agent and not a principal, to concentrate on the organisation and management of the construction operations. The project team, including the construction manager, is responsible for all financial administration associated with the works. The construction manager is paid an agreed fee to cover the costs of its staff and overheads. This is generally considered to be the least adversarial form of contract and is often invoked when design needs to run in parallel with construction.

Relevant issues

Variations from the formats described above can be a potential source of confusion and compromise the intended philosophies. Before a contractual or organisational variation is introduced the choice of procurement option should be made against the most important criteria. Only then should essential variations be introduced and these must be dealt with by specific contractual arrangements and documentation within the framework of the overall procurement method adopted.

It is important to recognise that the actual process by which construction projects are implemented remains identical whichever procurement route is followed. This process involves four stages:

1. Development of a detailed definition of the requirements for the producer (the detailed project brief).

2. Preparation of designs, working drawings and specifications identifying every component and detailing the construction method.

3. Procurement of every component required for the product, and the specialised skills necessary for its construction.

4. Management of the activities of the many participants and contributors involved in the project.

Construction components comprise those prefabricated or manufactured wholly or partially off-site, as well as those which are constructed on-site. There is an ongoing trend towards more use of components manufactured off-site, and towards the use of more proprietary components, which means that significant elements of design are carried out by the suppliers of the components, rather than by the design team directly responsible to the client. The implications of this on the design and the build responsibilities of the project as a whole must be carefully thought through and incorporated in the project handbook.

Characteristics of alternative procurement options

Table 3.1 provides a comparison of the features of the four basic procurement options available. The characteristics are explained in detail in Appendix 15.

Appointment of project team

The project manager in consultation with the client will decide on and implement a selection procedure for the members of the project team and may then appoint the project team on behalf of the client. Alternative methods of procurement will affect the selection procedure.

There are two main arrangements for a project team appointment:

■ separate appointment of independent service providers

■ single appointment of a team of service providers or a lead organisation for the provision of all services.

It is important that the members of the project team should be as compatible as possible both in temperament and in working methods, if the project is to have the greatest likelihood of success.

The project team may be selected and appointed through a process of short-listing and structured interviews or through a competitive tendering procedure. This may be through EU procurement procedures, which may be mandatory for publicly funded projects (dependent on size of project). (See Appendix 5 for an introduction to EU procurement rules.) The project manager must be fully informed on all issues related to the procurement process and advise the client accordingly.

The client should be consulted on the formulation of short lists and should be invited to attend any interviews. The process is set out in Table 3.2.

Table 3.1 Characteristics of alternative procurement options

	Characteristic	Traditional	Design and build	Management contracting	Construction management
1	Diversity of responsibility	Moderate	Limited	Large	Large
2	Size of market from which costs can be tested	Moderate	Limited	Moderate	Large
3	Timing of cost certainty	Moderate	Early	Late	Late
4	Need for early precise definition of client requirements	No	Yes	No	No
5	Availability of independent assistance in development of design brief	Yes	No	Yes	Yes
6	Speed of mobilisation	Slow	Fast	Fast	Fast
7	Flexibility in implementing changes	Reasonable	Limited	Reasonable	Good
8	Availability of recognised standard documentation	Yes	Yes	Yes	Limited
9	Ability to develop proposals progressively with limited and progressive commitment	Reasonable	Limited	Reasonable	Good
10	Cost-monitoring provision	Good	Poor	Reasonable	Good
11	Construction expertise input to design	Moderate	Good	Good	Good
12	Management of design production programme	Poor	Good	Good	Good
13	Influence in selection of trade contractors	Limited	None	Good	Good
14	Provision for monitoring quality of construction materials and workmanship	Moderate	Moderate	Moderate	Good
15	Opportunity for contractor to exploit cash flow	Yes	Yes	Yes	No
16	Financial incentive for contractor to manage effectively	Strong	Strong	Weak	Minimal
17	Propensity for confrontation	High	Moderate	Moderate	Minimal

For the short-list method the project manager should formulate the short lists, convene and chair the interviews, record and assess the results and present a report and recommendation to the client for final decision (see Appendix 14 and Part 2 Appendix B for further guidance on selection process).

Most professional firms are members of organisations that publish standard terms of appointment and codes of conduct. It is usual to appoint the project team members on the standard terms which are designed to provide a proper balance of risk and responsibility between the parties. Standard terms are capable of amendment by agreement but the project manager should advise against terms which impose uninsurable risks or unquantifiable costs on the consultants or are in conflict with their professional responsibilities or codes of conduct.

The project manager will issue to the appointed project team the project execution plan, project handbook, the outline project brief and the master schedule together with the budget or cost plan. It is advisable that these elements are referred to in as much detail as is available at the time of the project team's appointment.

Table 3.2 Characteristics of alternative procurement options

Activity	Considerations
■ **Selection and appointment of project manager**	May be appointed at inception/feasibility stage
■ **Agree criteria for team selection**	Type of expertise and scope Budget fee Contractual procurement strategy
■ **Define in detail each assignment**	Extent of services required Co-ordinate with other professional agreements
■ **Define roles and duties**	Scope of work Roles and duties
■ **Agree terms and conditions of engagement**	Client's or standard conditions of engagement Programme, PI insurance, warranties
■ **Selection of those invited to bid**	Utilise relevant databases Agree list of consultants Decide selection criteria Format proposal required Agree content and fee
■ **Carry out the selection process**	Agree interview team Agree information to be used Arrange interviews Utilise scoring system
■ **Negotiate conditions of appointment**	Provide client with analysis and selection recommendations Negotiate final conditions
■ **Advise on final appointment**	Issue letter of appointment Issue letters of rejection
■ **Oversee formalities relating to PI insurance, warranties and building defects insurance**	Legal department to formalise Finance department for fees Legal documents issued

Partnering

Partnering was identified in the Latham Report as a set of actions by project teams by which conflict could be minimised. The intention is to provide a 'win–win' situation for the partners. It is put into practice by having regular partnering workshops where all the key members of the project team work to establish and foster co-operative ways of working aimed at improving performance. In broad terms partnering teams agree mutual objectives that take account of the interests of all the parties; establish co-operative methods of decision making including procedures for resolving problems quickly; and identify actions to achieve specific improvements to normal performance. The workshops take place throughout the project initially under the guidance of an independent partnering facilitator.

It has already been shown that partnering can bring benefits in the form of reduced costs, improved quality and shortened timetables. It should be considered particularly for clients with rolling programmes or phased projects but it can provide benefits for every project. Ideally, partnering includes the supply chains that produce key elements (see Appendix 8). Overall partnering is part of the project management process.

4 Pre-construction stage

Client's objectives

At this stage the client expects to finalise the project brief with the project team, identify and agree the solution that gives optimum value, and to ensure a detailed design which can be efficiently delivered with predictability of cost, time and quality.

Interlinking with previous stages

After the client has made a commitment to the project, accepted the feasibility report and approved the scheme design, the process will then move into the next phase or stage which we call pre-construction.

However, it should be appreciated that many of the stages overlap and it is only to identify the full scope of activities involved in the development process and to enable some sort of chronology to be established, that we have separated the activities into stages.

'Pre-construction' involves establishing the detailed design, the preparation of tender documents and the tendering process (including negotiated tendering). However, the precise sequence of activities will depend very much on the choice of procurement system, and the type and form of contract selected.

It is worth noting at this stage, that we are moving into an ever-increasing legislative environment, with greater controls in the form of statutory requirements, national and European legislation and guidelines, minority stakeholder pressures, demand for greater sustainability and growing restrictions on disposal of unused material, to name but a few.

Therefore, by the start of this pre-construction stage, a significant number of key activities will have been addressed and action taken. These include the following:

■ The client's project brief detailing the project objectives will have been established and the associated scheme design fundamentally completed. However, while it would be expected that the detailed project brief would remain substantially unchanged for the remainder of the project, it is likely that unforeseen factors will have some effect upon the brief during the project period, although hopefully these will be minimal.

■ A suitable site and the scope of any treatments required will have been identified and made available.

■ Environmental and energy audits will have been undertaken.

■ Risk register prepared incorporating data from risk analysis.

■ Surveys to cover: geology, topography, hazardous materials (COMAH), landfill and recycling will have been carried out.

- Obligatory reports concerning sustainability, disability discrimination, etc., will have been prepared and approved by the appropriate authorities.

- Statutory requirements concerning the Housing Grants, Construction and Regeneration Act (see Appendix 7) and CDM regulations will have been accommodated.

- Statutory authorities, public bodies and utilities will have been approached for information regarding all mains services, highways and related infrastructure items, which are likely to influence site development.

- A master project schedule will have been prepared.

- A cost plan will have been prepared.

- A cost allowance will have been allocated to cover on-site development including enabling works, infrastructure, buildings, fitting-out and equipment.

- The planning authorities will have been contacted regarding the planning status of the site, which has been deemed acceptable for the intended purpose. Outline planning consent will have been obtained.

- The project team will have been appointed together with their associated consultants. This team will include the client, the project manager and, as soon as possible, representatives from the main contractor and associated key subcontractors/work packages. These will all contribute to the strategic decision-making process.

- The project execution plan drafted during the feasibility stage may be enhanced during this stage. It is a live document which governs the strategy, organisation, control procedures, respective responsibilities for the project and much more:

  - client brief: functional and aesthetic; business plan

  - constraints and risk assessment; revenue assumption/criteria

  - funding cost controls: budget; drawdown procedures; reserves

  - schedule: deadlines, milestones

  - organisation and resources: responsibilities, delegated authority

  - project strategy and procurement details

  - roles and responsibilities of project team members

  - occupation plan: commissioning; facilities management/maintenance strategy.

The project handbook would have been prepared under the guidance of the project manager and submitted to the client and any other interested party, for comment, discussion and agreement. Its review and update will be the responsibility of the project manager, unlike the 'health and safety file' which is the responsibility of the planning supervisor. This project handbook differs from the PEP, in that the handbook sets out the process and procedures for administration purposes, whereas the PEP covers detail as shown in Chapter 2 and in the preceding bullet points.

The client will have authorised the project to proceed and should be aware that considerable costs will be incurred. Adequate cash flow provision must be provided for regular monthly expenditures. These will include professional services fees, e.g. for the project manager, architect, quantity surveyor, structural and M&E engineers, together with planning fees and on-site investigations, demolition, site clearance and disposal, etc.

The pre-construction stage is about final preparation for the construction stage, the success of which will depend, to a great extent, on the amount of planning and preparation that has taken place during this and earlier stages.

Design management

The project manager will need to convene a meeting of the design team and any other consultants/advisers to review all aspects of the project to date. A dossier of relevant information should be circulated in advance. The object of the meeting will be to formulate a design management plan.

The plan should at least cover:

- who does what by when

- the size and format of drawing types

- schedules of drawings to be produced by each discipline/specialist

- relationships of interdependent CAD (computer-aided design) systems

- transfer of data by information technology

- estimates of staff hours to be spent by designers on each element or drawing

- monitoring of design resources expended compared to planned estimates

- schedules of information required/release dates

- initiating procedures for design changes

- incorporation within the design schedule of key dates for review of design performance to check:

  - compliance with brief

  - cost acceptance

  - value engineering analysis

  - health and safety issues

  - completeness for tender.

The project manager, as a basis for monitoring and controlling the design process, will use the agreed design management plan. While the project manager may convene a meeting of the design team, responsibility for the co-ordination and integration of the work of other consultant and specialists lies with the design team leader. For certain elements of the project different lead consultants will be nominated. However these roles will always come under the direction of the design team leader for co-ordination.

Suggested task list for design team leader

- Establishing the overall design style, quality, etc.

- Establishing a grid/reference system for the base scheme.

- Reviewing the design schedule.

- Directing the design process.

- Liaising with the client about significant design issues.

- Preparing sufficient production information for consultants and specialists to develop their proposals, co-ordinating these and integrating them into the overall scheme.

- Advising on the need for and appointment of other consultant and specialists.

- Establishing a system for information transfer, check compatibility of system and software.

- Co-ordinating the briefing document.

- Establishing a system of design reviews and validation.

- Agreeing a basis for the cost plan to be developed and subsequent monitoring.

- Advising the client of their role and duties under the CDM regulations.

Duties of project manager at this stage

- Organising within the client organisation appropriate groups of people, who will contribute to the detail of the brief and champion relevant aspects of the design prepared by the design team for signing off.

- Assisting in the preparation or finalisation of the detailed project brief.

- Preparing the design management plan.

- Arranging the appointment of other consultants and specialists.

- Organising the communication and information systems.

- Producing co-ordinated design schedule and monitoring progress.

- Ensuring that various technical specialists appointed by the client such as IT, acoustics, catering, landscaping and artists are brought into the design process at the appropriate times.

Project co-ordination and progress meetings

To aid control of the design process the project manager will arrange and convene project progress meetings at relevant intervals to review progress on all aspects of the project and initiate action by appropriate parties to ensure that the design management plan is adhered to. Distributing minutes of meetings to all concerned is an essential part of the follow-up action.

Design team meetings

Design team meetings are convened, chaired and minuted by the design team leader. It is not essential for the project manager to attend all these meetings as a matter of course, although he normally has the right to do so. The project manager will receive minutes of all meetings and will report to the client accordingly.

Managing consultants' activities

Key specialist contractors need to be involved early and managed equally with the consultants (see Figure 4.1).

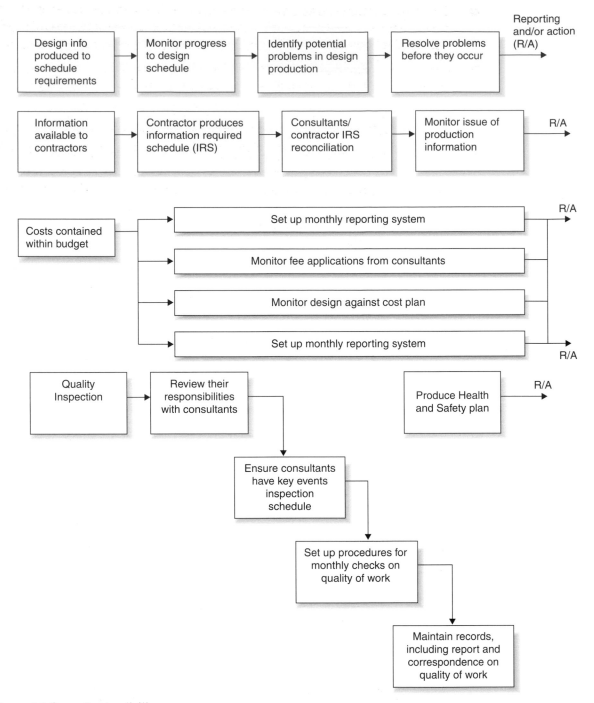

Figure 4.1 Consultant activities

The project manager has several responsibilities:

- Monitoring progress against the design management plan in association with the team. This is essential in view of their interrelationship. However, effective interrelationship cannot be finalised until the full team has been appointed and has had time to get to grips with the project and its complexities.

- Advising the design team leader of the requirement to agree the detail and integration of the design team activities and to submit an integrated design production schedule for co-ordination by the project manager.

- Incorporating, into the project schedule, dates for the submission of design reports and periods for their consideration and approval.

■ Commissioning, as necessary, or arranging for the team to commission, specialist reports, e.g. relating to the site, legal opinions on easements and restrictions and similar matters.

■ Ensuring a competent consultant is appointed as planning supervisor as required by CDM regulations.

■ Drawing to the attention of the client and the designers their respective duties under the CDM regulations and monitoring compliance.

■ Arranging for the team to be provided with all the information they require from the client in order to execute their duties. It is an important function of the project manager to co-ordinate the activities of the various (and sometimes numerous) participants in the total process. Planning supervisor, solicitors, accountants, tax advisers, development advisers, insurance brokers and others may all be involved in the pre-construction stage.

■ Submitting, in conjunction with the design team leader, preliminary design proposals, reports and scheme design drawings to the client for approval (see Figure 4.2).

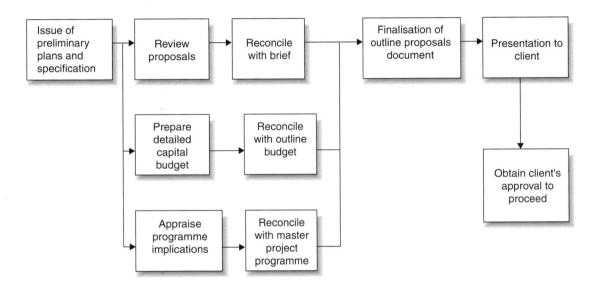

Figure 4.2 Outline design proposals

■ Conveying approvals to the team to proceed to subsequent stages of the project.

■ Obtaining regular financial/cost reports and monitoring against budget/cost plans. Initiating remedial action within the agreed brief if the cost reports show that the budget is likely to be exceeded. Solutions to problems that cannot be resolved within the agreed brief, or likely substantial budget underspend, should be submitted to the client with recommendations. The necessity to agree firm budgets at an early stage is most essential. It could, in certain cases, lead to the client modifying the project brief.

■ Preparing 'Schedule of Consents' with action dates, submission documents, status, etc., and monitoring progress.

■ Checking that professional indemnity insurance policies are in place and remain renewed on terms that accord with conditions of engagement.

Statutory consents

Although a great deal of the detailed work involved in obtaining statutory consents, such as planning permission and Building Regulations approval, is carried out by the design team and other consultants, the project manager has a vital facilitating role to play in what can be critical project activities.

Planning approval

Legislation

The primary legislation governing the planning process is contained in the following Acts of Parliament:

■ Town and Country Planning Act 1990

■ Planning (Listed Buildings and Conservation Areas) Act 1990

■ Planning (Hazardous Substances) Act 1990

■ Planning and Compensation Act 1991

■ Town and Country Planning (General Permitted Development) Order 1995

■ Town and Country Planning (General Development Procedure) Order 1995

■ Town and Country Planning (Use Classes) Order 1987

■ Town and Country Planning (Development Plan) Regulations 1991.

The grant of planning approval does not remove the need to obtain any other consents that may be necessary, nor does it imply that such consents will necessarily be forthcoming.

Planning permission

Planning permission is required for any development of land. 'Development' is defined in section 55 of the 1990 Act as 'the carrying out of building, engineering, mining or other operations in, on, over or under land, or the making of any material change in the use of any buildings or other land'. The definition of building operations includes the demolition of buildings.

Section 55 also provides that certain works and uses do not constitute development under the 1990 Act. These include:

■ works of maintenance, improvement of other alteration of any building which affect only the interior of a building or which do not materially affect its external appearance

■ the use of buildings or land within the curtilage of a dwelling house for any purpose incidental to the enjoyment of the dwelling house such as the use of land for the purpose of agriculture or forestry; and

■ change of use of land or buildings from one use to another within the same class of the Town and Country Planning (Use Classes) Order 1987.

Moreover, the General Permitted Development Order 1995 grants permissions for certain defined classes of development, mainly of a minor character. The most commonly used class permits a wide range of small extensions or alterations to dwelling houses.

Schemes for Enterprise Zones and Simplified Planning Zones (see PPG5 – Planning Policy Guidance 5) also grant planning permission for developments for types defined in the scheme concerned.

Timing

Planning permission cannot be guaranteed or assured in advance of the local planning authority (LPA) decision and the project manager must recognise this in the master schedule.

Negotiations The project manager will normally assist the design team leader in negotiations with officers of the local authority and report to the client on the implications of any special conditions, or the need to provide planning gain through the appropriate statutory agreements. The client's legal advisers are briefed to act for the client accordingly.

Presentations The project manager will arrange, should it be necessary, any presentations to be made to LPAs and local community groups. He or she will also organise meetings, including agreeing publicity and press releases with the client.

Refusal Should planning permission be refused, the advice of the relevant consultants should be obtained and action initiated, either to submit amended proposals or to appeal the decision.

Appeal In the event of an appeal, arrangements are made for the appointment and briefing of specialists and lawyers, including managing the progress of the appeal. Applicants who are refused planning permission by an LPA, or who are granted permission subject to conditions which they find unacceptable, or who do not have their applications determined within the appropriate period, may appeal to the Secretary of State. Appeals are sent to the Planning Inspectorate.

Enforcement powers The authority's main enforcement powers are:

■ to issue an enforcement notice, stating the required steps to remedy an alleged breach within a time limit (there is a right of appeal to the Secretary of State against a notice)

■ to serve a stop notice which can prohibit, almost immediately, any activity to which the accompanying enforcement notice relates (there is no right of appeal to the Secretary of State)

■ to serve a breach condition notice if there is a failure to comply with a condition imposed on a grant of planning permission

■ to apply to the High Court or County Court for an injunction to restrain an actual or apprehended breach of planning control

■ to enter privately owned land for enforcement purposes; and

■ following the landowner's default, to enter land and carry out the remedial work required by an enforcement notice, and to charge the owner for the costs incurred in doing so.

After an enforcement notice has become effective, or at any time after a stop notice has been served, it is a criminal offence not to comply with an enforcement notice's requirements or to contravene the prohibition in a stop notice.

Other statutory consents

It is the duty of the design team to facilitate that the design complies with all other statutory controls, e.g. consents for Building Regulations, means of escape, the storage of hazardous materials, fumes and emissions, and pollutants. Generally, statutory controls make the owner or occupier responsible for the aspect of continuing duties in relation to the statute. The project manager obtains all consents from the design team and/or other relevant sources, and arranges for the client to be advised of these continuing duties. Others such as specialist subcontractors submit and obtain Building Regulations approval for their product/system.

Detail design and production information

The project manager's monitoring and co-ordinating role will entail extensive liaison with members of the project team and will include the tasks shown in Figure 4.3, which are set out in more detail below.

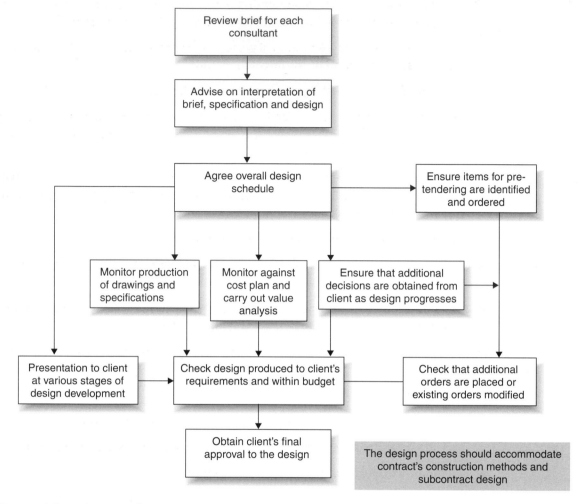

Figure 4.3 Co-ordination of design work up to design freeze

- ■ Controlling the extent to which the design will be produced by specialist contractors and/or component manufacturers, and establishing the division of responsibilities between them and the design team.

- ■ Reviewing the project strategy, control systems and procedures, and amending the project handbook, as required.

- ■ Amplifying the design brief as necessary during design development.

- ■ In conjunction with the project team, preparing a detailed stage schedule for the detailed design and production information stage, defining tasks and allocating responsibilities.

- ■ Preparing schedules to establish timely flow of information from the design team for:

    - ○ cost checking

    - ○ client's approval

- ○ tender preparations

- ○ construction processes.

- ■ Co-ordinating the activities of the client and the project team in the management of the production of the design information.

- ■ Formulating, in collaboration with the consultants, recommendations to the client/owner in respect of the quality control system, including:

  - ○ on- and off-site inspection of work for compliance with specifications, and testing of materials and workmanship

  - ○ performance testing and the criteria to be used

  - ○ preparation of schedules for required samples and mock-ups, updating and monitoring progress of approvals; copies of schedules are included in the relevant monthly reports.

- ■ Listing the key criteria in terms of performance benchmarking that in all areas of design make clear how the design will be judged, i.e. any changes or faults with current facilities.

- ■ Monitoring the emerging detail design against the cost plan.

- ■ Liaising with the client/project team and the local authority/utilities and other statutory bodies to obtain permissions and approvals.

- ■ Evaluating changes in client's requirements for cost and time implications and incorporating approved items into the design process.

- ■ Monitoring progress and providing regular reports incorporating information relating to:

  - ○ project status

  - ○ progress against schedule, together with exceptions report

  - ○ cost against budget/cost plan, together with reconciliation statement

  - ○ forecast of total cost and date of completion

  - ○ critical areas

  - ○ corrective action needed.

- ■ Obtaining the client's approval to the detailed design and production information phase.

- ■ Initiating arrangements for implementation of approved design and production information, to ensure that contractors' reasonable information requirements are fulfilled.

Tender action

The procurement schedule will indicate the time allowed for short listing suitable contractors or works packages (manufacturers/suppliers and installers of large tender items such as 'cladding' and early items such as groundworks and foundations). This schedule will also show activities such as tender interviews, tendering and selection. It will then lead to a design schedule which defines scope, release dates, approval periods, cost checking and consolidation into tender documentation.

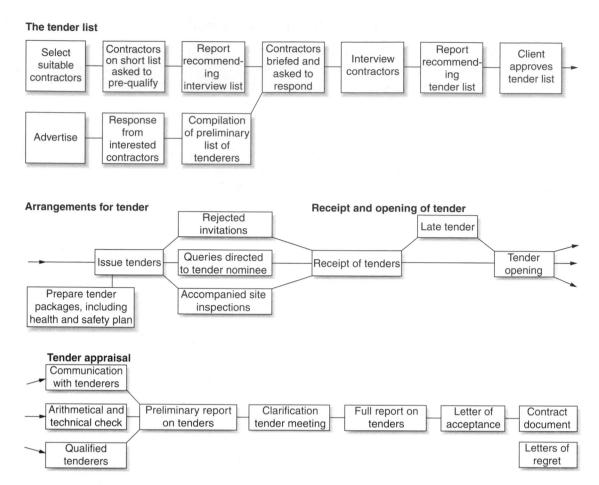

Figure 4.4 Tender procedure. (Note: this figure reflects the traditional approach. The tendering process will vary depending on the procurement route chosen.)

These activities may include the following:

- Checking that the various tender documents are produced at appropriate times, including those for enabling works (e.g. demolition, site clearance, access and hoarding) and ensuring that they contain any special terms required by the client. In conjunction with the relevant consultants, preparing lists of firms to be invited to tender for the main and subcontract elements of the work (pre-qualifying process). Obtaining confirmation that the listed firms will be prepared to submit tenders at the specified dates, taking up references and/or interviewing prospective tenderers, together with the relevant consultants.

- Ensuring that appropriate reference to the CDM regulations is made in tender documentation where the contractor is to be appointed as 'principal contractor', including the health and safety plan.

- Checking, in liaison with other project team members, that all subcontract terms are compatible with the main contract terms, paying particular regard to contractor-designed elements and confirming that appropriate warranties are secured. Receiving reports on tenders, together with method statements. Interviewing successful tenderers, if necessary, to clarify any special conditions and to meet significant leading personnel. Arranging for formal acceptance of tender as appropriate and issuing relevant letters of intent.

- Selection should be based on balancing quality and price (see CIB and CIRIA publications on selection process for further guidance on balancing quality and price in selection process).

■   Initiating action if tenders are outside budget.

■   Ensuring that the client understands the nature and terms of the construction contract, particularly those in relation to possession and payment terms, and that possession of the site can be given to the contractor on the date set out in the tender.

■   Arranging for formal signing and exchange of contracts.

In view of the EU directives on procurement, negotiated tendering may also be undertaken as an option to secure best value for money. For further information about the negotiated tendering process see Appendix 5.

Bringing the contractor on board

It is always advisable to ensure that the contractor is brought onto the design team and design stages from the earliest possible opportunity as is viable. This concept is of course dependent on the type of contract under which the project is to be executed.

The benefits gained by early involvement of the contractor are (the encouragement to ensure the acceptance and involvement of the contractor is solely dependent on the positive management of the project manager):

■   Resolution of buildability issues at design stage.

■   Choice of the most efficient materials to be used.

■   Advice to the client as to costing (practical, rather than rates).

■   Opportunity to involve specialist subcontractors in the design process to streamline and pinpoint design issues.

■   The contractor can understand the client's needs in all areas, which will assist in the construction process, leading to improved quality and other benefits for the client.

The contractor can highlight any health and safety issues which must be taken into account on the design.

Pre-start meeting

The pre-start meeting with contractors and consultants (project team) is held to establish proper working arrangements, roles and responsibilities, and lines of communication, and to agree procedures to be followed throughout the contract (project on site). If bonds are required they must be provided before possession of site is granted. The 'principal contractor's' health and safety plan must be in place before work starts on-site.

See Table 4.1 for a specimen agenda for a pre-start meeting.

Table 4.1 Specimen agenda for pre-start meeting

1. **Introductions**
   Appointments, personal
   Roles and responsibilities
   Project description

2. **Contract**
   Priorities
   Handover of production information
   Commencement and completion dates
   Insurances
   Bonds (if applicable)
   Standards and quality

3. **Contractor's matters**
   Possession
   Schedule
   Health and Safety files and plan
   Site organisations, facilities and planning
   Security and protection
   Site restrictions
   Contractor's quality control policy and procedures
   Subcontractors and suppliers
   Statutory undertakers
   Overhead and underground services
   Temporary services
   Signboards

4. **Resident engineer/Architect/Clerk of works' matters**
   Roles and duties
   Facilities
   Liaison
   Instructions

5. **Consultants' matters**
   Structural
   Mechanical
   Electrical
   Others

6. **Quantity surveyor's matters**
   Adjustments to tender figures
   Valuation procedures
   Remeasurement
   VAT

7. **Communications and procedures**
   Information requirements
   Distribution of information
   Valid instructions
   Lines of communication
   Dealing with queries
   Building Control notices
   Notices to adjoining owners/occupiers

8. **Meetings**
   Pattern and proceedings
   Status of minutes
   Distribution of minutes

Agenda items at pre-start meeting

Introduction

■ Introduce the representatives who will regularly attend progress meetings and clarify their roles and responsibilities. The client, contractor and consultants may wish to introduce themselves.

■ Briefly describe the project and its priorities and objectives, and any separate contract that may be relevant (preliminary, client's own contractors, etc.).

■ Indicate any specialists appointed by the client, e.g. for quality control, commissioning, for this contract.

Contract

■ Describe the position with regard to preparation and signature of documents

■ Hand over any outstanding production information, including nomination instructions, variation instructions. Review situation for issuing other important information.

■ Request that insurance documents be available for inspection immediately, remind the contractor to check specialist subcontractors' indemnities. Check whether further instructions are needed for special cover.

■ Confirm the existence, status and use of the information release schedule, if used. Establish procedure for agreeing adjustments to the schedule should they be necessary.

Contractor's matters

■ Check that the contractor's master schedule is in the form required and that it satisfactorily accommodates the specialist subcontractors. It must:

   ○ contain adequate separate work elements to measure their progress and integration with services installations

   ○ allocate specific dates for specialist subcontract works, including supply of information, site operations, testing and commissioning

   ○ accommodate public utilities, etc.

■ Agree a procedure for the contractor to inform the architect of information required in addition to any shown on the information release schedule. This is likely to involve the contractor's schedule of information required, which must relate to his works schedule and must be kept up to date and regularly reviewed. It should include information, data, drawings, etc., to be supplied by the contractor/specialist subcontractors to the architects/consultants.

■ Review in detail the particular provisions in the contract concerning site access, organisation, facilities, restrictions, services, etc., to ensure that no queries remain outstanding. Ensure the contractor has a copy of any conditions placed on the client in respect of the planning consent. Also provide the contracts with legal drawings showing the curtilage of the site ownership.

■ Quality control is the contractor's responsibility. Remind the contractor of the contractual duty to supervise standards and quality of work during the execution of the works.

■ Numerous other matters may need special coverage, e.g.:

   ○ check whether immediate action may be needed by the contractor over specialist subcontractors and suppliers

- emphasise that drawings, data, etc., received from contractor or specialist subcontractors (not approved) will remain the responsibility of the originator

- review outstanding requirements for information to or from the contractor in connection with specialist works;

- clarify that the contractor is responsible for co-ordinating the performance of specialist works and for their workmanship and materials, for providing specialists with working facilities, and for co-ordinating site dimensions and tolerances.

■ The contractor must also provide for competent testing and commissioning of services as set out in the contract documents, and should be reminded that the time allocated for commissioning is not a contingency period for the main contract works.

■ The contractor must obtain the architect's written consent before subletting any work.

Resident engineer/ architect/clerk of works' matters

■ Clarify that architect's inspections are periodic visits to meet the contractor's supervisory staff, plus spot visits.

■ Explain the supportive nature of the various roles and the need for co-operation to enable them to carry out their duties.

■ Remind the contractor that the resident staff must be provided with adequate facilities and access, together with information about site staff, equipment and operations.

■ Confirm procedures for checking quality control, e.g. through:

- design and methodology

- certificates, vouchers, etc., as required

- sample material to be submitted

- samples of workmanship to be submitted prior to work commencing

- test procedures set out in the bills of quantities

- adequate protection and storage

- visits to suppliers'/manufacturers' works.

Consultant's matters

■ Emphasise that consultants will liaise with specialist subcontractors only through the contractor. Instructions are to be issued only by the architect/ contract administrator. The contractor is responsible for managing and co-ordinating specialist subcontractors.

■ Establish working arrangements for specialists' drawings and data for evaluation (especially services) to suitable timetables. Aim to agree procedures which will speed up the process; this sector of work frequently causes serious delay or disruption.

Quantity surveyor's matters

■ Agree procedures for valuations; these may have to meet particular dates set by the client to ensure that certificates can be honoured.

Clarify:

- that dayworks will only be accepted on written instructions

- that daywork sheets are required within a stated number of days from work being carried out

- tax procedure concerning VAT and 'contractor' status

- that the contractor should only order from drawings and specifications, not the bills of quantities.

Communications ■ The supply and flow of information will depend on a schedule being estab-
and procedures lished at the start and will proceed smoothly if:

- there is regular monitoring of the information schedules

- requests for further information are made specifically in writing, not by telephone

- the design team responds quickly to queries

- technical queries are raised with the clerk of works (if appointed) in the first instance

- policy queries are directed to the architect/contract administrator

- discrepancies are referred to the architect/contract administrator for resolution.

■ On receiving instructions, the contractor should check for discrepancies with existing documents; check that documents being used are current.

■ Information to or from specialist subcontractors or suppliers must be via the contractor.

■ All information issued by the design team is to be via the appropriate forms, certificates, notifications, etc. The contractor should be encouraged to use standard formats and classifications.

■ All forms must show the distribution intended; agree numbers of copies of drawings and instructions required by all recipients.

■ Clarify that no instructions from the client or consultants can be accepted by the contractor or any subcontractor; only empowered written instructions by the architect/contract administrator are valid and all oral instructions must be confirmed in writing. Explain the relevant procedure under the contract. The contractor should promptly notify the architect/contract administrator of any written confirmation outstanding.

■ Procedures for notices, applications or claims of any kind are to be strictly in accordance with the terms of the contract; all such events should be raised immediately the relevant conditions occur or become evident.

Meetings

Review format, procedures, timing, participants and objectives of the next stage:

- meetings, i.e. site (progress) meetings, policy/principal's meetings and contractor's production information meetings, and

- site inspections.

Fee payments

The project manager is responsible, as required in his or her conditions of engagement, for receiving fee accounts and invoices from consultants and others concerned with the project, checking for correctness and arranging for payment within the terms of the various appointments or contracts.

Quality management

It is the project manager's role to set up and implement an appropriate process to manage project quality. From the quality policy defined in the project brief, the development of a quality strategy should lead to a quality plan setting out the parameters for the designers and for the appointment of contractors. Quality control then becomes the responsibility of the contractor, subcontractors and suppliers operating within the agreed quality plan. The plan itself should establish the type and extent of independent quality auditing (particularly for off-site production of components) and the timing of inspections and procedures for 'signing off' completed work.

It is the responsibility of the design team and other relevant consultants to specify the goods, materials and services to be incorporated in the project, using the relevant British Standards, codes of practice and Agreement Board criteria or other appropriate standards.

The achievement of these standards rests with the appointed main contractor. When interviewing contractors at pre-tender stage, the project manager will seek confirmation that each company has a positive policy towards the control of quality, a policy which will be reflected in all of its operations on-site.

Dispute resolution

Although it is hoped that the non-adversarial approach advocated in the Latham Report and the increasing choice of alternative procurement options and partnering will lead to a reduction in disputes, nevertheless, the project manager should make every effort to pre-empt any dispute that may arise and endeavour to mitigate and resolve the problem.

Another option which may prove to be an effective alternative to arbitration, adjudication or litigation, subject to the conditions of the contract, is the application of alternative dispute resolution (ADR); it provides:

- a viable method of early resolution to avoid claims and likely high costs

- a conciliation service at short notice

- a third party free of any conflict or interest; this generally leads to a negotiated settlement.

An outline of procedures to be applied in resolving contractual disputes is given in Appendix 16.

5 Construction stage

Client's objectives

At this stage the client should aim to ensure the safe completion of the construction/development within the targets set at previous stages.

Interlinking with previous stages

The change from pre-construction to the construction stage signifies the culmination of all the pre-construction effort that allows the actual work to start on-site. With this change, the duties of the project manager also change and this section sets out his or her tasks on the premise that he or she then supervises the construction and final delivery of the project.

This move to construction must be managed as a seamless operation and must recognise and enact any key policy or strategy decisions that will have been taken during the earlier stages of the project cycle. Decisions will have been taken in such areas as client's main requirements, planning requirements, whole life-cycle constraints, value engineering, procurement methods, early contractor or specialised subcontractor/supplier involvement, health, safety and welfare, environmental issues, etc. The procedures and responsibilities for all these have been dealt with in the earlier sections of this *Code of Practice* and now must be effectively implemented during this dynamic stage of the project.

This does not mean however, that there is no opportunity left for further initiative or development of the project. On the contrary, a very proactive input is needed from the project manager and all the members of the project team to search out and find further practical betterment that will enhance the end product.

It is the project manager's overriding role during this phase to provide the team with the necessary strong and proactive leadership. He or she has to steer the project to completion through continuous measurement of performance against time, quality and costs and to carry out all necessary actions to ensure the team's successful delivery of a project that not only satisfies the client, but also exceeds expectations.

Responsibilities of the project manager at this stage

To be the proactive 'driver' of the project

The project manager needs to demonstrate his soft skills as well as hard skills:

- *Hard skills* generally include planning, scheduling, organisational ability, report writing, information assembly, cost control, innovation, decision making and prioritisation.

■   *Soft skills* include leadership, motivation, communication, interpersonal skills, personality, team-building abilities, honesty, integrity and sense of humour.

To set the project objectives

The project manager has responsibility for defining the primary objectives for the project. From these he or she has to develop individual objectives, team objectives and the project general objectives, in order to achieve the primary objectives. The project manager has then to be able to communicate effectively with the team members and obtain their commitment to achieve the objectives.

These should include:

■   meeting the client's objectives within the contract, i.e. not at all costs

■   fair treatment of all parties to the project

■   customer focus

■   ensuring a delighted client.

To ensure achievement of objectives

The project manager must clearly adhere to the project success criteria. He or she must maintain the measurement against the progress and then pro-actively manage the project to ensure success.

Achieving client's satisfaction

This has to be the prime responsibility of the project manager.

Roles of project team members

Although the precise contractual obligations of the project participants vary with the procurement option adopted, the project participants must carry out certain essential fundamental functions.

Client   Usually a client would have a relatively nominal direct involvement in the construction works; their chief interest would be:

■   to satisfy themselves that the contractor(s) were performing in accordance with the contract

■   to make sure they are meeting their obligations to pay all monies certified for payments to the consultants and the contractor(s).

Project manager The project manager has a role which is principally that of monitoring the performance of the main contractor and the progress of the works, and involves the following activities (some of which may have been accomplished in the pre-construction stage):

■   Ensuring contract documents are prepared and issued to the contractor.

■   Ensuring the contracts are signed.

■   Arranging the handover of the site from the client to the contractor.

■   Reviewing contractor's construction schedule and method statements.

- Ensuring procedures are in place and being followed.

- Ensuring site meetings are held and documented.

- Monitoring construction cash flow.

- Reviewing progress with contractor.

- Monitoring performance of contractor.

- Ensuring the health and safety file is being maintained.

- Ensuring design information required by contractor is supplied by consultants.

- Establishing control systems for time, cost and quality.

- Ensuring site inspections are taking place.

- Confirming insurance cover on the works.

- Managing project cost plan.

- Ensuring client meets contractual obligations (i.e. payments).

- Reporting to client.

- Managing introduction of changes.

- Ensuring statutory approvals are being obtained.

- Ensuring all relevant legal documents are in place.

- Collateral warranties:

  o   performance bonds

  o   reviewing construction risks.

- Establishing mechanisms for dealing with any claims.

- Anticipating and resolving potential problems before they develop.

Design team   The design consultants are responsible for:

- providing production information (i.e. details of building components)

- approving working drawings being provided by specialist contractors

- responding to site queries raised by contractors

- inspecting the works to ensure compliance with the drawings and specification

- inspecting the works to ensure an acceptable quality standard has been achieved.

The structural engineer will have a general duty of care to ensure the erection of the structural frame is proceeding in a safe manner. This might be extended to cladding fixings and other architectural components which are subjected to stress, forces or loadings.

Most building contracts refer to a contract administrator, usually the design team leader or the project manager, who is the formal point of contact between the project team and the contractor, and who has a contractual obligation in relation to the issuing of formal instructions to the contractor; these include:

- issuing design information

- issuing variations

- instructions on standards of work and working methods

- arbitrating on contractual issues

- issuing interim payment and other certificates

- issuing practical completion certificate.

Quantity surveyor

The quantity surveyor has a duty to:

- measure the value of work executed by the main contractor

- agree monthly valuations with the main contractor

- agree the final account with the main contractor.

The quantity surveyor has a separate responsibility to the client, usually through the project manager, for reporting on the overall financial aspects of the project.

Main/principal management contractor

The main/principal management contractor has responsibility for:

- mobilising all labour, subcontractors, materials, equipment and plant in order to execute the construction works in accordance with the contract documents

- ensuring the works are carried out in a safe manner

- indemnifying those working on the site and members of the public against the consequences of any injury resulting from the works.

Construction manager

A client may decide on a construction management route, directly employing a construction manager as a consultant acting as an agent and not a principal, with expertise in the procurement and supervision of construction. In this arrangement the construction manager's role is:

- to determine how the construction works should best be split into packages

- to produce detailed construction schedules

- to determine when packages need to be procured

- to manage the procurement process

- to manage the overall site facilities:

  - access

  - storage

  - welfare

- to supervise the package contractor's execution of the works.

In the management contracting arrangement, a management contractor acting as a principal would have the additional direct contractual responsibility for the performance of the package contractors.

Subcontractors and suppliers

Subcontractors have specialist expertise, usually trade related (e.g. mechanical or electrical installations, lift installation, joinery, demolition), for the supply and installation of an element of the total works.

Subcontractors may be either nominated or named by the consultants or selected and appointed directly by the main contractor, known as domestic subcontractors. If nominated, the client carries some risk in respect of the subcontractor's performance.

Suppliers provide certain materials, components or equipment for others to install.

Labour-only subcontractors provide only labour to carry out the installation of materials, components or equipment provided by the main contractor (e.g. carpenters, bricklayers and plasterers).

Due to their specialist knowledge, subcontractors have an increasing design responsibility for the detailed design related to their installations (may include fixing details, fabrication details, co-ordination with other installations).

There is a general obligation on all the project team to ensure the site is a safe working environment, although legally this falls to the principal contractor under CDM regulations.

Other parties   A large number of other bodies will be involved during the course of the construction works, these include:

■ **Building control officer** (local authority or approved inspector)
Inspection of various elements of the works (e.g. foundations, structure).

■ **Highways authority**
Inspection and adoption of roads and sewers.

■ **Environmental health officer**
Inspections related to pollution (e.g. mud, noise, smoke, water)
Also inspection of certain installations (e.g. drainage, kitchens).

■ **Fire officer**
Inspection of site for fire escape and hazards, storage of certain materials
Inspection of protection systems.

■ **Health and safety executive**
Inspection of site for safety aspects.

■ **Planning officers**
Checks on compliance with consents
Inspections of preserved trees.

■ **Archaeologists**
Inspection of excavations for ground disturbance.

■ **Trade unions**
Meetings with members in relation to complaints about site conditions.

■ **Landlord's representatives**
Inspections of scope and quality of works.

■ **Funder's representatives**
Inspections of progress and quality of works in order to release money.

■ **Police**
Discussions on traffic control, unloading and complaints.

Team building   Traditionally it has always been easy to execute contracts and projects using the contract and the specific duties and responsibilities for each party. However, this rigid approach has more often than not brought about an adversarial environment between usually the contractor and the client's design team.

Construction is a people business and thus communication is the key to a successful project. This involves the project manager heading up the professional design and construction team (including the contractor) and building trust between all the parties. The project manager will be responsible for the ultimate outcome of the project and thus it is in his or her interest that there is a united team working towards the same goal. There are numerous individual methods of team building which can be adopted by referring to the large literature published on management. However, the most effective and the critical period for the project manager to drive home his or her stance on the team is at the outset of the project, as and when designers and consultants together with the contractor are brought on board.

Regular progress meetings and workshops (both formal and informal) can assist in developing the bond between all members of this team. More importantly, during the construction stage, the team must have a hands-on (immediate) approach in resolving, assisting and thus eliminating any issues hindering the smooth progress of construction.

Construction is not about individuals but teams, both at pre-construction stage and during the work.

Health, safety and welfare
Construction (Design and Management) Regulations 1994 is the primary legislation dictating the duties, roles and responsibilities for key members in a construction project team.

The purpose of the CDM regulations is to identify the key individuals within the construction project. Specific duties are placed on:

- client
- designers
- planning supervisor
- principal contractor.

The project manager has a duty to monitor the activities and actions being executed by the above four individuals. This does not mean that the project manager will be held liable for any wrong-doing, but simply that the management of the project from the inception stage through design, construction and finally occupation, must have health and safety firmly in mind, eliminating any risks at each stage.

There are number of tools for monitoring the successful process of health and safety, the main ones being risk assessments, risk workshops, method statement analyses and the health and safety file (before and during construction) of individual potentially 'risky' design solutions and on-site construction method issues.

During the pre-construction stage the responsibility for health and safety is mainly on the design team and the client, and thus the management process must be considered by the project manager. During construction, the principal contractor is responsible for safety and welfare on-site. If the standard of method statements concerning activities is not to the satisfaction of the planning supervisor or the project manager, further meetings and discussions must be arranged between parties to agree an amicable safe way forward. The principal contractor is responsible for preparing a construction health and safety plan. This is constantly updated. At completion of the project the client is handed a concise health and safety file on the built product highlighting any potential risks to the end user.

Health, safety and welfare is the responsibility of *all* individuals involved in the construction. The project manager must take an active role in monitoring the process. He or she *must* emphasise the importance of health and safety considerations to, most importantly, the client and also the design and construction team. (See Appendix 2 for further information on health and safety including CDM regulations.)

Environment management systems

Environmental statements

Environmental concerns will increasingly affect projects especially with the pressure to develop brownfield sites and reuse old sites. The cost of addressing contaminants or other environmental issues can add significant costs and extend the project duration.

Planning authorities are also more likely to instruct environment studies (ES) and impose restraints as part of the planning process all of which must be incorporated into the project during the construction stage.

It is the project manager who has overall responsibility to ensure compliance with these aims, objectives and constraints.

The project manager will need to:

- Understand and act on environmental impact studies (see Appendix 11).

- Ensure proper environmental advice is available.

- Ensure the contractor is complying with the ES criteria.

- Seek and ensure action by the contractor of any remedial work should it be necessary to comply.

Contractor's environment management systems

The contractor must establish environmental management systems (EMS), but it is for the project manager to ensure that it is being managed properly and is progressing sufficiently to achieve all ES objectives.

Therefore, the project manager should:

- Receive details of the contractor's EMS and environmental plan (EP) specific to the project.

- Ensure that the contractor has set up all necessary procedures and structure to manage the EMS and implement the objectives of the EP.

- Check that the contractor's environment management plan matches the aims and objectives of the ES.

- Agree with the contractor any further aims, specific targets or initiatives that will maximise sustainability of the project and minimise the detrimental impact of the construction process.

- Proactively monitor the progress of the contractor to maintain environmental objectives.

Contractual arrangements

The project manager has to ensure that all statutory and contractual formalities are in place prior to allowing work to start on site. It may mean that he has to ensure that others have given the relevant notice, and if appropriate, received the relevant approval. A log may help to keep track of notices and approvals together with the owner of the task.

These are likely to include:

- planning: architect

- CDM notification: planning supervisor

- third-party insurance: contractor

- Public indemnity (PI) insurance: consultants

- Notice to start work under the Building Regulations: contractor

- fire regulation compliance: architect

- performance bonds: contractors.

Also on completion various completion certificates are required, these should be specified in the particular specification, and would include:

- fire regulation compliance

- electrical completion certificate

- test certificates for both manufacturing and installation

- lifting beams, tests and marking

- Building Regulation compliance

- pressure vessel and boiler certificates.

For special buildings or processes, particular licences and certificates may be required; for example, nuclear projects, pharmaceutical, oil and gas, rail. If there is any doubt ask the design team for their advice, then manage the process.

Establish site   Once the design has been finalised and contracts have been signed, the project is ready to go on-site. It is imperative that the site set-up process is carried out and completed in the most efficient manner prior to the start of the main construction works. The issues that the project manager must be aware of and monitor with the contractor at this stage are not only practical and physical operations but also administrative plans and procedures agreed by the parties. The areas where the project manager is to agree and monitor site set-up are:

- Ensure that site boundaries are clearly identified with the contractor.

- Establish the contractor's proposal for security.

- Establish the contractor's proposal for emergency plans in case of fire or any incidents.

- Establish the contractor's proposal for site accommodation, specifically the suitability of the welfare facilities.

- Carry out a survey of existing conditions of the site and the adjacent properties. Record any relevant issues in liaison with adjoining owners where possible.

- Establish with the contractor the administrative procedures such as daily returns, daily diaries, faxes, e-mail facility, drawing issues, etc. This activity is most important as it will set out the communication route between all parties throughout the project. The findings and agreements with the contractors should be recorded by the project manager and distributed to all professionals involved.

- Ensure that the contractor is aware of and is attending to any issues that may be present due to neighbours being close to the site including the terms of any party wall awards or rights of light issues.

- Ensure that the contractor has clearly identified the health and safety risks that exist on the site.

- Ensure that all signage is displayed correctly.

The above issues are to be agreed with the contractor. The project manager can not dictate how the contractor is to set up site. The project manager role must be advisory and thus monitor that the correct actions as agreed are being implemented.

LIVERPOOL JOHN MOORES UNIVERSITY
Aldham Robarts L.R.C.
TEL. 051 231 3701/3634

Control and monitoring systems

It is the project manager's prime duty to make sure that all necessary control and monitoring systems are properly set up and implemented by the contractor.

The project manager must ensure that these systems produce the most appropriate information and reports, on a regular and timely basis, so that he or she can use them to monitor and manage the project to a successful conclusion.

By carrying out his or her own audits and checks of the systems, the project manager must be fully satisfied with the accuracy of the data produced and that they do indeed indicate the 'real' position at any point in time and, where appropriate, accurately forecast the final position of the project.

Such contractor's systems will generally be (but not limited to):

■ quality management system

■ schedule management system

■ quality control system

■ cost monitoring and management system

■ health, safety and welfare system

■ environmental management system

■ document management system.

It is of absolute importance that the project manager fully understands the relevance of the information being produced by these systems. The project manager must proactively use this information to manage the contractor and the project team through regular management meetings. The aim is not only to understand where the project is and where it is ultimately going, but also to identify any potential problem areas at a sufficiently early stage so that any rectification procedures and/or mitigation measures can be taken to ensure best delivery of the project.

Contractor's schedule

The project manager has a duty to the client to monitor the performance of the contractor. In order to carry this out adequately the project manager needs to ensure the contractor has prepared a construction schedule in sufficient detail to enable the construction works to be closely monitored.

The project manager needs to receive and review the contractor's schedule prior to the commencement of the works in order to:

■ check that it complies with the client's time requirements

■ check that it acknowledges any restraints imposed on the construction of the works

■ ensure that the level of detail is appropriate for the complexity of the works

■ ensure that it is suitable for monitoring the progress of the works

■ confirm the sequencing and logic of the schedule.

The construction schedule must be supported by an information requirement schedule that realistically informs the project manager when outstanding design information is required in order for the contractor to achieve the schedule dates.

Regular reports recording progress achieved against the schedule must be received from the contractor, and a progress status agreed with the contractor.

Any rescheduling of the works necessary to recover delay situations needs to be received, reviewed and agreed.

In addition to the detailed analysis of progress, the project manager should examine high-level progress trends to obtain an overall view of project status. This can involve graphically comparing cumulative planned progress against actual achieved.

The contract usually requires that the contractor prepare a contract schedule that becomes part of the contract. This schedule does not normally go into a great deal of detail, as time-scales, dependencies and interfaces have yet to be agreed with subcontractors to the main contractor.

The project manager will need to obtain a working schedule and more detailed schedules to cover specific sections of work. These might include fitting out schedules, phased handover, commissioning, and a completion schedule showing how completion will be achieved.

Rescheduling required as a result of changes or slippages will be required to show how time can be recovered or the effect on the completion dates.

It is the duty of the project manager not only to monitor the contractor's progress, but also to monitor any work being undertaken by other advisors, suppliers or companies that have an independent input into the completion of the project. These should all be monitored against the overall client's master schedule with its own milestone and targets. The project manager is managing the overall project for the client and its successful delivery.

Value engineering (related to construction methods)

Value engineering (VE) is an exercise that involves most of the project team as the project develops, by selecting the most cost-effective solution. However, VE is about taking a wider view and looking at the selection of materials, plant, equipment and processes to see whether a more cost-effective solution exists that will achieve the same project objectives.

VE should start at project inception where benefits can be greatest, however the contractor may have a significant contribution to make as long as the changes required to the contract do not affect the time-scales or completion dates, or incur additional costs that outweigh the savings on offer.

There is, however still a place for VE, especially at the start of construction. The application of the job plan (see Table 5.1) remains consistent but the detail available is obviously more than during the design and pre-design stages.

The 'result accelerators' originally proposed by Miles still act as useful guides to VE at the construction stage (see Table 5.2).

In all of this it is most important to remember the relationship between cost and value – value is function divided by cost.

Concentration on the function of the project or product will avoid mere cost-cutting.

Table 5.1 Value engineering job plan

Information
Function analysis
Speculation
Evaluation
Development
Recommendation
Implementation

Table 5.2 Result accelerators

Avoid generalities
Get all available costs
Use information from best source
Blast, create and refine
Be creative
Identify and overcome road blocks
Use industry experts
Price key tolerances
Use standard products
Use (and pay for) specialist advice
Use specialist processes

The project manager must take a proactive role in both giving direction and leadership in the VE process, but must above all ensure that time and effort are not wasted and do not have a detrimental effect on the progress of the project.

An example of a VM framework has been included in Appendix 10.

Management of the supply chain The contractor has overall responsibility for the management of his supply chain to meet his contractual obligations.

The project manager has the duty to ensure that this chain is being effectively managed so as to avoid any potential delay, unnecessary cost implications or any other adverse effect on the delivery of the project.

This is an important issue as it is so very often the case that problems further down the contractual chain can be responsible for long delays and/or major disputes right back up through the whole chain. They can potentially result in the deterioration of relationships with the contractor and have a knock-on effect with not only his performance, but also the performance of the project team as a whole.

Duties and responsibilities should include:

■ Receive and understand the details of the contractor's supply chain and the controls to manage it.

■ Establish key members and linkages within the chain.

■ Receive and interrogate reports from the contractor about the ongoing progress, including any reports from procurement managers and expeditors.

■ Implement a regular monitoring system to check the progress of key suppliers or subcontractors (against contractor's delivery schedule) so that timely warning signals flag-up any potential delays or failures that could have an adverse effect on the progress and financial stability of the project.

■ Agree with contractor any appropriate remedial action that may be needed to rectify any problem areas.

Risk register The risk register (see Appendix 9) is a document that should be prepared at the earliest stages of the project, identifying potential risks throughout the project. This register should be reviewed and updated according to circumstances and stages of the project. At the construction stage the risk register should be reviewed to include any new construction risks.

In addition to monitoring those construction-related risks previously identified in the project-wide risk register, the project manager needs to ensure the contractor has instigated a risk management system for those risks likely to impact on the actual construction works.

The project manager needs to ensure that the contractor:

■ establishes a fully detailed listing of construction risks

■ determines the likely probability and impact of each risk

■ reviews the risks with the project team

■ prepares method statements and action plans demonstrating how risks will be mitigated or managed out

■ identifies, and informs, the person responsible for managing each risk

■ prepares contingency plans for any key risks having a significant impact

■ regularly reviews and reports on the status of risks.

Benchmarking In certain circumstances, particularly when framework or partnering agreements are in place, it may be appropriate to employ benchmarking of a contractor's performance against the best industry practice.

A major difficulty of benchmarking in construction is locating base data that allow for meaningful comparisons. Since 1998, as part of the annual production of statistical information gathered from contractors, the government has collected measures of key performance indicators. These provide the most widely sourced comparators currently available to benchmark individual companies against the industry average levels of performance.

A number of construction clients commission their own research to assemble, from other similar organisations, meaningful performance data that allow them to carry out benchmarking of the companies they use.

Benchmarking is closely associated with the concept of continuous improvement, and a company's performance can be monitored over time to confirm that improvement measures introduced are effective.

Change and variation control The project manager should carry out the following tasks to control variations:

■ Variations which result from changes to the project brief (to be avoided whenever possible) (see Figure 5.1) or design/schedule modifications (e.g. client's request, architect's or site instructions) must follow a procedure which:

  ○ identifies all consequences of the variation involved

  ○ takes account of the relevant contractual provisions

  ○ defines a cost limit, above which the client must be consulted and, similarly, when specifications or completion dates are affected

  ○ authorises all variations only through a change order system initiated by the project manager (see Figure 5.1 for changes to the client's brief, and Appendix 16 for an example of a change order pro forma).

■ Identifying, in consultation with the project team, actual or potential problems and providing solutions which are within the time and cost limits and do not compromise the client's requirements, with whom solutions are discussed and approval is obtained.

■ Checking the receipt of scheduled and/or ad hoc reports, information and progress data from project team members.

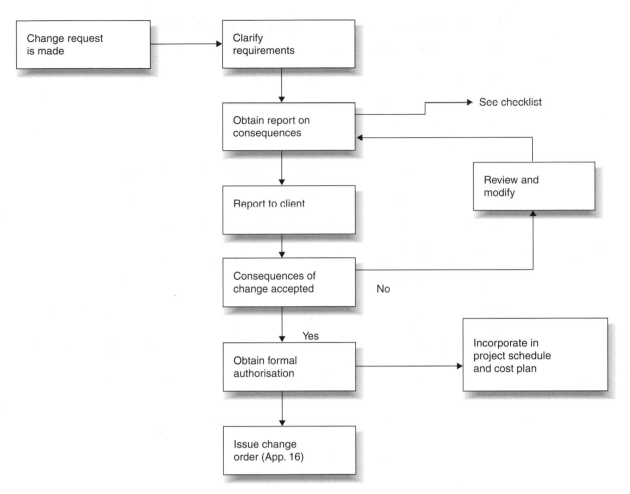

Figure 5.1 Changes in the client's brief

Table 5.3 Changes in the client's brief: checklist

Activity	Action by
1. Request for change received from client	Project manager
2. Client's need clarified and documented	Project manager
3. Details conveyed to project team	Project manager
4. Review of technical and health and safety implications	Consultants and project manager
5. Assessment of schedule implications	Planning support staff and project manager
6. Evaluation/calculation of cost implications	Quantity surveyor
7. Preparation of report on effect of change	Project manager in consultation with consultants
8. Reporting to client	Project manager
9. Consequences accepted/not accepted by client	Project manager
10. Non-acceptance – further review/considerations as per items 4, 5, 6 and action items 7, 8	Project manager assisted by consultants
11. Further reporting to and negotiation of final outcome with client	Project manager assisted by consultants
12. Agreement reached and formal authorisation obtained	Project manager
13. Incorporation into project schedule and cost plan (budget)	Project manager and quantity surveyor
14. Change order issued (see Appendix 16)	Project manager and client

The main way of minimising claims or variations is to ensure the brief is clearly defined, and that the contract documents and drawings accurately and completely reflect the detail.

Managing change control at the design development stage is far more effective than managing the process during construction.

Circumstance-driven changes, mistakes or unknowns have to be effectively managed on the basis that in many instances time is more expensive than the material change. Some form of authorisation needs to be agreed (financial limits) so that instructions can be given without having to refer every change back to the customer for approval.

The project manager will need to maintain a register of changes and variations, cross-referenced to the contractor's request-for-instruction notices, and possible contract claims. The register should include budget costs and final costs for reporting to the client on a regular basis.

Accurately detailed daily diaries will need to be kept, complete with plant, labour and material deliveries so that consequential costs can be identified.

In dealing with the effects and costs of variations, the project manager will need, where possible, to agree costs before issuing an instruction. It is also wise to agree, again where possible, that work will be undertaken with no overall effect on the schedule.

It is vital to carefully record events and the situation at the time.

Procedurally, the project manager must inform both the design consultants and the main contractor that all variation instructions must be in the correct written form and *must* only be issued via the project manager unless he or she is the appointed contract administrator in the main contract. To avoid unnecessary complications in agreeing valuations and accounts, it is imperative that the variation instructions are issued from one source. Design consultants must raise (in writing) requests for instructions, to the project manager, who will in turn issue the instructions to the contractor. *All* variations *must* have an instruction against them in order to be valued.

Supervision of the works

Once the project is underway on-site regular inspections and monitoring of progress are to be carried out by the project manager. There is a fine line as to how involved the project manager should get with the everyday issues facing the contractor, and thus the relationship as mentioned previously will determine the appropriate approach.

It is the project manager's responsibility to arrange from the outset progress meetings usually at 2–3-week intervals. During these meetings the contractor will present a report as to progress on-site with any relevant design issues which require resolving. Note that these meetings are set up to monitor progress and not to cover technical design. If necessary separate design meetings should also be arranged. The reporting process to the project manager must not be restricted to the contractor but must also include all designers and consultants. It is at these forums that the project manager must manage and ensure all parties are working together and achieving individual target dates for producing information and maintaining progress against schedule.

Not withstanding the formal progress meetings, if not resident, the project manager should visit site at least twice during the week and spend limited time on-site discussing progress with site staff and chasing up the appropriate individuals for information and progress.

Reporting   A fundamental aspect of the project management role is the regular reporting of the current status of the project to the client.

The project manager needs to ensure an adequate reporting structure and calendar are in place with the consultants and contractors. Frequency and dates of project meetings need to be co-ordinated with the reporting structure.

Reporting is required for a number of reasons:

■ to keep the client informed of project status

■ to confirm that the necessary management controls are being operated by the project team

■ to provide a discipline and structure for the team

■ as a communication mechanism for keeping the whole team up to date

■ to provide an auditable trail of actions and decisions.

Progress reporting should record the status of the project at a particular date against what the position should have been; it should cover all aspects of the project, identify problems and decisions taken or required, and predict the outcome of the project.

The project manager needs to receive individual reports from the consultants and contractor and summarise them for the report to the client. The detailed reports should be appended as a record.

Typical contents of a project manager's project report would contain:

■ executive summary

■ legal agreements

■ design status

■ planning/Building Regulations status

■ procurement status

■ construction status

■ statutory consents and approvals

■ project schedule and progress

■ project financial report

■ variation register update

■ risk register update

■ major decisions and approvals required.

Trends shown visually are an excellent mode of conveying information to clients and senior management.

Public liaison   The client would probably have set out his overall public relations and liaison strat-
and profile   egy during the pre-construction stages of the project.

In reflection of this, the project manager should take a leading role in 'local' public relations during the construction stage. This will improve the public's perception of the construction industry in general.

Such activities or actions should include:

■ Ensuring no local nuisance or negative impact arising from the project.

■ Maintaining good housekeeping both on-site and in the immediate off-site area.

■ Erecting informative scheme boards and public viewing platforms.

■ Ensuring the contractor takes part in a local or national 'Considerate Contractor' scheme.

■ Taking awareness initiatives with local schools.

■ Attending local public meetings to raise the profile of the project.

■ Organising site visits for local schools, residents and business people.

■ Partaking in local environmental schemes or issues.

■ Being involved with fundraising for local charities or causes.

Commissioning and operating and maintenance (O&M) manuals

Commissioning The main commissioning and putting to work issues are covered in Chapters 6 and 8 – this section covers the construction stages.

The project manager should receive within the early stages of construction the contractor's commissioning schedule in order to be satisfied that it is properly co-ordinated with the building works schedule. (As an example, the balancing of the heating and air conditioning system can only take place when the building envelope and internal spaces have been secured.)

A problem may occur in that in many instances the building services contractor is a subcontractor to the main contractor and the subcontract may not be in place at this early stage of construction. In this case, the main contractor will need to identify the logic and sequence of the commissioning.

Operating and maintenance manuals The production of the O&M manuals is covered under the engineering services commissioning section. With systems becoming ever more complex a properly user-friendly set of manuals is important. Detailed requirements should be set out in the contract, so that there is no dispute on the level of detail required.

The CDM regulations now cover O&M manuals, and it is the planning supervisor's role to ensure that they are delivered as part of the health and safety file. These manuals should also include details of the complete building with input from all the design team. The project manager has to monitor the progress that the planning supervisor is making on assembling these files and if needs be, ensure that all necessary actions are taken to expedite their completion.

Payment A vital part of the construction process is ensuring the contractor and subcontractors receive regular payment for the work which has been carried out.

The project manager has a role in attempting to avoid disruption caused by contractors failing on the project. There are a number of actions the project manager can take:

■ checking on the financial standing of contractors prior to their appointment

■ ongoing monitoring of the financial position of contractors

■ ensuring all payments due are promptly paid by the client.

Generally the contractor makes a monthly application for payment. The quantity surveyor values work, the architect or contract administrator certifies it and the client pays for it within a stipulated period of the application/certification date.

The project manager has an important role in ensuring the client honours obligations to pay contractors against certificates authorised by the contract administrator.

As projects become larger and more complex so do the means of finance. These include:

■ public private partnership (PPP)

■ public finance initiative (PFI)

■ design build finance operate (DBFO)

■ build own operate transfer (BOOT)

■ plus - cost plus, reimbursable, target cost, cost plus fee.

These forms of contract are more likely to have their own 'tailored' methods and formats for payments to the contractor or concessionaire, but set out below are the more common methods of payment for traditional or design and build contracts.

■ **Valuations**
The traditional method of payment has been a physical measuring of the works carried out on-site and the quantity of work costed against the rates in the bill of quantities. Carried out jointly by the main contractor and quantity surveyor, this is done monthly. The contract administrator issues an interim certificate for the amount due to the client and the client has to make payment to the main contractor within a period stated in the contract.

■ **Milestones**
Tenderers as part of their tenders are asked to break down their total price into a number of sums against pre-determined milestones. Milestones are usually the completion of elements of the construction works (e.g. completion of the structure up to a certain level). Normally likely to be 20–40 milestones.

Acknowledgement that a milestone has been achieved by the contract administrator will release payment of the sum to the contractor. This can sometimes be called an 'Activity Schedule' (NEC contract).

■ **Stage**
Similar to milestone payment but likely to be far fewer stages (e.g. completion of superstructure, achievement of watertight building).

■ **Earned value**
Regular payments made in accordance with an earned value system. Payment will be related to actual progress position achieved on the works. As the value of payments is based on the schedule assessment of progress it avoids the need to conduct separately monthly measurements of works carried out.

■ **Ex-gratis**
Although not a formally recognised method of payment, in certain extreme cases when lack of cash is preventing a contractor from carrying out obligations under the contract, a special one-off payment may be made with the client's agreement. This is an ex-gratis payment made in advance of the normal payment procedure to ensure certain works are carried out in order to recover or prevent a delay situation or to expedite certain materials. In cases this might be accompanied by a pre-payment insurance bond. It is important if payment is for materials or equipment that ownership is clearly established to guard against insolvency of the contractor.

Construction completion report (including lessons learnt)

The project manager should compile a construction completion report on the completion of the construction stage of the project.

This report should include:

- An analysis of the performance of the contractor.

- An analysis of the performance of the project team, including all advisors.

- Performance of original quality, cost and time targets for the project against those finally achieved.

- Performance of the project against its benchmarking criteria and all other targets and objectives that have been monitored.

- A detailed *lessons learnt* analysis.

For the sake of good construction practice, it is extremely important that the lessons learnt are identified, understood and noted down. It is best done while events are fresh in the memory and before the team has been disbanded. There then needs to be an appropriate management procedure for transferring these in order that others (the client, host company or other participants to the project) might learn for future projects.

6 Engineering services commissioning stage

Client's objectives

At this stage the client should be satisfied that the engineering installation has been installed correctly, in a safe manner, and that it performs to the requirements of the design.

The project manager's objective is to ensure that the commissioning of the separate systems is properly planned and executed, so that the installation as a whole is fully operational at handover without delay to the schedule and that any fine tuning necessary after handover is carried out in liaison with the client/user.

Interlinking with construction

It must be stressed that the location of this chapter does not mean that the activities involved only take place at the end of the construction stage. Engineering commissioning is a very important part of the construction process and must be addressed and considered very early within the project. The following are suggested activities that must be considered well before this stage:

- Decide the most appropriate time within the project to appoint the commissioning contractor and the role/scope of work.

- Where appropriate, appoint a commissioning contractor to review the design drawings and working drawings to ensure commissionability.

- Ensure consultants clearly identify testing and commissioning requirements.

- Ensure consultants/client identify performance/environmental testing requirements.

- Ensure that the project schedule includes sufficient time to undertake the specified commissioning, and in particular, the additional time required for any performance/environmental testing and statutory testing to authorities.

- Clearly identify the method of presenting, recording and electronically storing 'as installed' information.

- Although not strictly part of engineering commissioning, ensure that the requirement for specialist maintenance contracts for equipment is carefully considered prior to awarding tenders for such equipment.

Commissioning generally

Commissioning is carried out in four or sometimes five distinct parts: (a) static testing of engineering services, (b) dynamic testing of engineering services,

(c) performance testing of engineering services (not always undertaken), (d) undertaking statutory tests for various authorities, and (e) client commissioning. Note that *performance* testing also includes *environmental* testing.

The first four items, *engineering services testing, commissioning, performance testing and statutory* tests, are part of the construction design and installation phases of the project. Client commissioning is an activity predominantly carried out by the client's personnel assisted, where required, by the consultants. This is dealt with in Chapter 7.

The engineering services testing and commissioning process objectives and main tasks are as described within this chapter.

Procurement of commissioning services

Smaller projects There are many ways to procure the commissioning specialist. On smaller projects, via the main contractor, the mechanical and electrical subcontractors are most likely to be responsible for the testing and commissioning of their installation. Electrical contractors will normally use in-house resources, except where specialist items of equipment require the manufacturer to assist with their testing. Mechanical contractors will usually appoint a *commissioning specialist* to work on their behalf. Again, where specialist items of equipment are installed, the mechanical contractor will request the manufacturer to assist with its testing where appropriate. However, it should be noted that often these *commissioning specialists* are no more than balancing engineers. This is fine for simple installations, but where more complicated systems are involved or specific commissioning and performance tests are required, their management and execution may not be adequate. Careful specification of the requirements within the design documentation is required when tendering the installation work. This is all too often ignored or given insufficient time and effort which inevitably creates problems later in the construction process.

Larger projects On larger projects the method of procuring the commissioning specialist can take many forms. In traditional forms of contract it can again be via the main contractor/ services contractor, however, in construction management or similar forms of contract, a specialist commissioning contractor is often appointed. This commissioning contractor normally fulfils one of two roles: the role of managing the testing and commissioning process (the actual work being done by the installation contractors as detailed for small projects above), or the role of undertaking the commissioning work. In this latter role, the point of delineation for testing/ commissioning between the installation contractor and the commissioning contractor is usually at the end of static testing and the start of dynamic testing. See below for a definition of these terms. This latter role is gaining in popularity for the following reasons:

- It provides a degree of independence to the commissioning process.

- The commissioning contractor is under the control of the construction manager/managing contractor and reports directly to them, giving greater control and transparency to the process.

In either role, the benefit to the project is that the commissioning contractor can be brought in to the project very early to manage the whole testing and commissioning process.

Role of the commissioning contractor

Below are some of the activities that can be included within the scope of work for the commissioning contractor:

- Review the design drawings near the end of design to ensure familiarity with the design intent and to add their expertise in to the commissionability of a scheme.

- Ensure that the testing and commissioning are correctly specified in the tender documentation.

- Review the services contractors' working drawings for commissionability.

- Set up the testing and commissioning documentation to create consistency between the various contractors.

- Define the method, media type, style and content of the 'as installed' information to create consistency between the various contractors.

- Manage the specialist equipment manufacturers' tests.

- Liaise with building control and other organisations for witnessing of relevant statutory tests (including insurers' tests).

All of these functions are often given insufficient thought on projects, so if they are not to form part of the commissioning contractor's brief, then it should be recognised that some other part of the project team should undertake this work.

The testing and commissioning process and its scheduling

Flow charts relating to the various stages of testing, *commissioning and performance testing* are given in Figures 6.1 and 6.2. It is important for the project manager to understand the differences between the terms testing, *commissioning and performance testing*, and to ensure that the schedule has sufficient time within it to enable these activities to be undertaken. Unfortunately, with this stage of the project being so close to handover, there is often pressure to gain time by shortening the testing, commissioning and performance/environmental testing schedule. This should be strongly resisted. Rarely, if ever, after the project will such an opportunity exist to test fully the services to ensure that they work individually, as a system, and that they work under part load and full load conditions. Many problems in respect of the under-performance of services within an occupied building can be related back to either insufficient quality in the testing and commissioning, or insufficient time to test and commission.

It should also be borne in mind that various statutory services will need to be demonstrated to building control (or the relevant government department if a Crown building, e.g. DCMS) and insurers. Time should be allowed for within the schedule since these activities are often taken as separate tests after the main commissioning has been undertaken.

Differences between testing and commissioning

Testing

During the services installation various tests will be undertaken known as 'static testing'. This testing is normally undertaken to prove the quality and workmanship of the installation. Such work is undertaken before a certificate is issued to 'liven' services whether electrically or otherwise. Examples of such are:

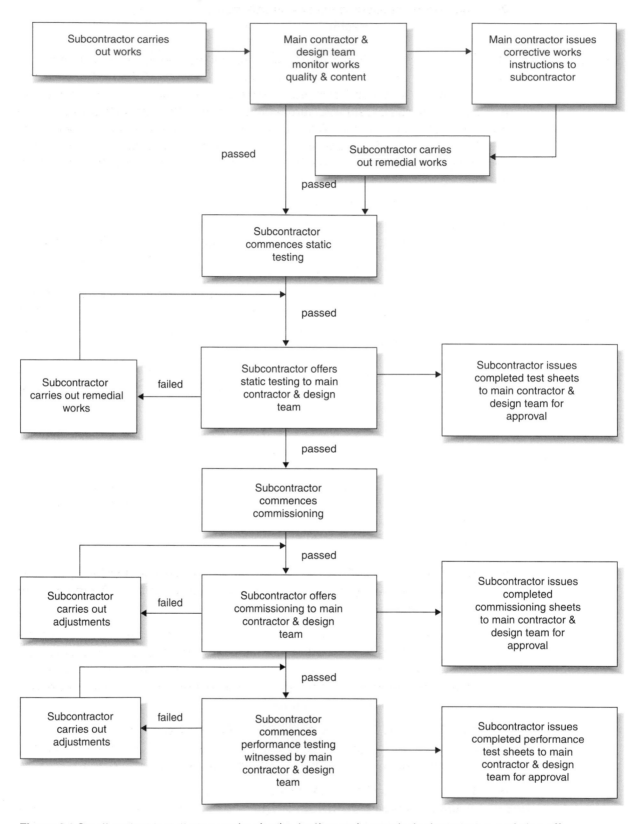

Figure 6.1 Small project installation works checks, testing and commissioning process and sign off

■ pressure testing ductwork and pipework

■ undertaking resistance checks on cabling.

Commissioning On completion of the static testing, dynamic testing commences, this being the commissioning. Commissioning is undertaken to prove that the systems operate and perform to the design intent and specification. This work is extensive and normally commences by issuing a certificate permitting the installation to be made 'live', i.e. electrical power on. After initial tests of phase rotation on the electrical installation and checking fan/pump rotation (in the correct direction), the more recognised commissioning activities of balancing, volume testing, load bank testing, etc., begin.

Performance testing On completion of the commissioning, performance testing can commence. Some may not distinguish between commissioning and performance testing. However, for scheduling purposes it is worth distinguishing between commissioning plant as individual systems and undertaking tests of all plant systems together, known as performance testing (and including environmental testing). Sometimes this performance testing is undertaken once the client has occupied the facility, e.g. for the first year, because systems are dependent on different weather conditions. In such cases, arrangements for contractor access after handover to fine-tune the services in response to changing demands must be made.

However, for some facilities it is desirable, if not demanded, to simulate the various conditions expected to prove that the plant systems and controls operate prior to handover, e.g. computer rooms.

Main tasks to be undertaken

To assist the project manager the following has been provided to summarise the main tasks to be carried out during the three main stages of pre-construction, construction and post-construction.

Pre-construction The following items will need to be confirmed:

■ The consultants/client recognise engineering services commissioning as a distinct phase in the construction process which has an important interface with client commissioning (see Chapter 7).

■ The relevant consultants identify all services to be commissioned and define the responsibility split for commissioning between designers, contractor, manufacturer and client. Responsibility for specialised plant/services is defined early, particularly 'wear and tear' and the cost of consumables, fuel, power, water, etc.

■ The services designers, and commissioning contractor if relevant, audit the final layout drawings to ensure that they make provision for the systems to be commissioned in accordance with the relevant codes of practice.

■ The consultants/client, and commissioning contractor if relevant, identify all required statutory and insurance approvals relating to services commissioning, and see that plans are made for meeting requirements and obtaining the approvals (see Appendices C and D of Part 2).

■ The client understands the importance of the presence of their own maintenance/engineering department/maintenance contractor during the commissioning process.

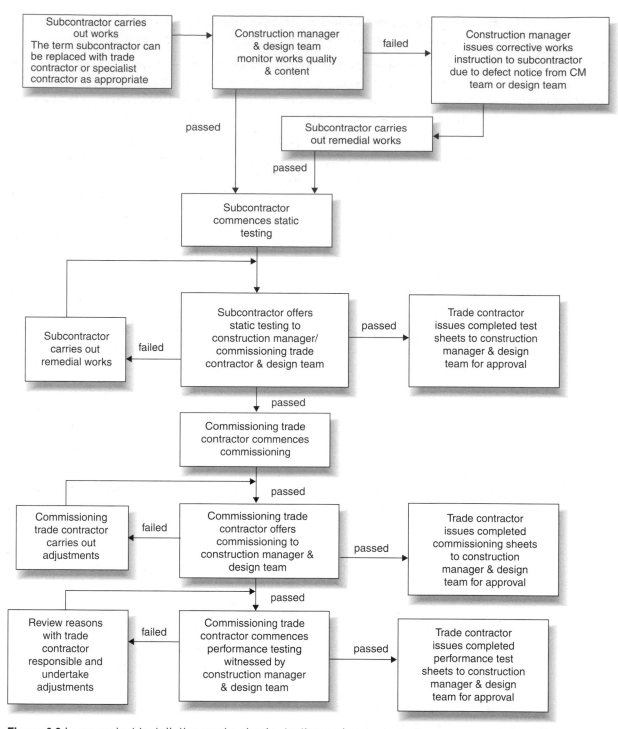

Figure 6.2 Large project installation works checks, testing and commissioning process and sign off

- The client considers whether an aftercare engineer needs to be appointed to support the client/user in the first 6–12 months of occupancy.

- There is a schedule showing the time-scale and sequence of commissioning and testing and handover events, system by system; this is essential.

- Arrangements are made to ensure that one person only is responsible for control and management of the client's role in commissioning of services. This could be the client's commissioning officer or the project manager, who should be a member of the client's team defined in Chapter 7. This does not preclude more than one person having the benefit of witnessing the commissioning process.

- The contract documents *must* make adequate provision for testing, commissioning and performance testing (see Appendices C and D of Part 2).

Construction and post-construction

- The consultants must inspect the work for which they have design responsibility, and report on progress and compliance with contract provisions, highlighting any corrective action necessary. A commissioning management specialist may be appointed to carry out much of this work.

- There must be confirmation that all the contractor's construction schedules include commissioning activities and that they are properly related to preceding construction activities. Activities must be complete, timings reasonable and compatible with planned handover, and properly related to preceding activities.

- Co-ordination of the consultants' arrangements is required for client involvement in or observation of contractors' commissioning against contract arrangements.

- Monitoring and reporting progress of commissioning will be carried out to ensure that activities start as scheduled and that the requirements for completion before handover are met. Corrective action will have to be initiated as necessary. It is important that commissioning activity durations do not become eroded due to late or incomplete construction work.

- All 'completed construction' documents should be in place before commissioning an individual system commences, e.g. cleaning out, testing the electrical power and controls to it. Also, the requirements of 'permits to work', health and safety at work should be met; and responsibility for insurance should be clearly defined.

- Statutory/insurance tests should be arranged and undertaken, witnessed by the relevant authority, e.g. building control, DCMS, utility companies, fire brigade, insurers.

- Commissioning records, e.g. test results, calibration requirements, certificates and checklists, must be properly maintained and copies bound into the operating and maintenance (O&M) manuals or in separate commissioning manuals to form part of the official handover documentation.

- O&M manuals, 'as installed' record drawings and client's staff training have to be provided by the contractor as required under the contract, although it is recommended that these are fully co-ordinated by others, e.g. the commissioning contractor, if appointed.

- Adopting agreed structure and software for O&M manuals with copy disks provided for ease of updating.

- Record drawings being provided in CAD format for ease of updating.

- Using video recordings during client training sessions for subsequent repeat visual reference and to assist new maintenance staff advancing along the learning curve.

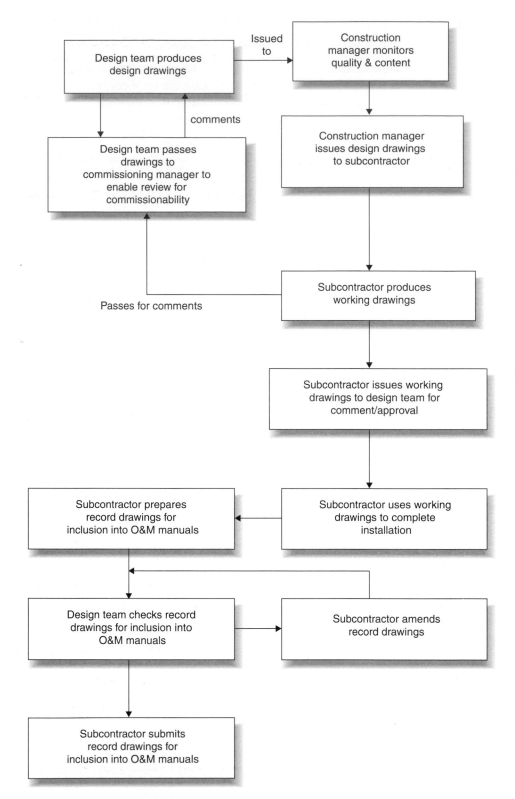

Figure 6.3 Project drawing issue flow chart

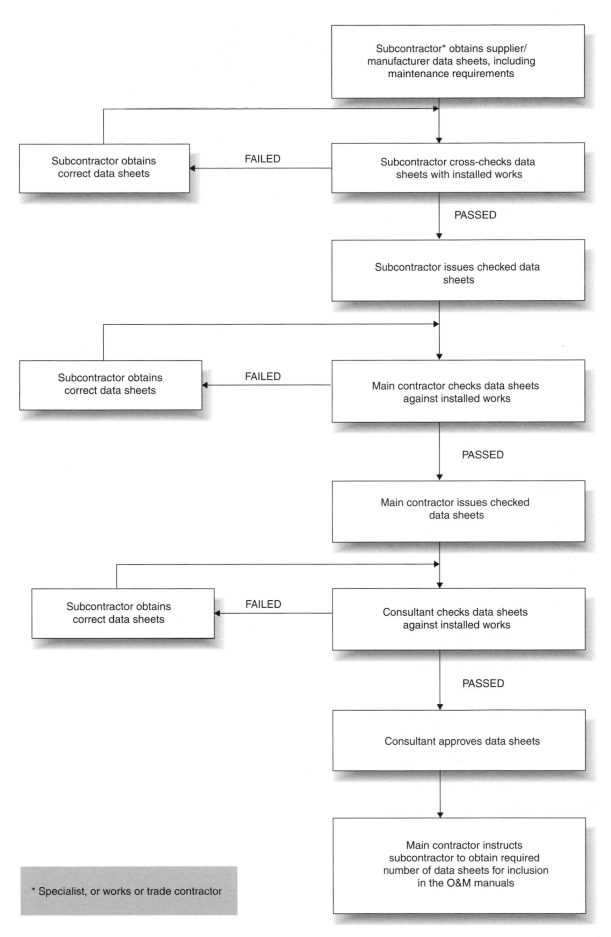

Figure 6.4 Services installation, testing and commissioning data sheets flow chart

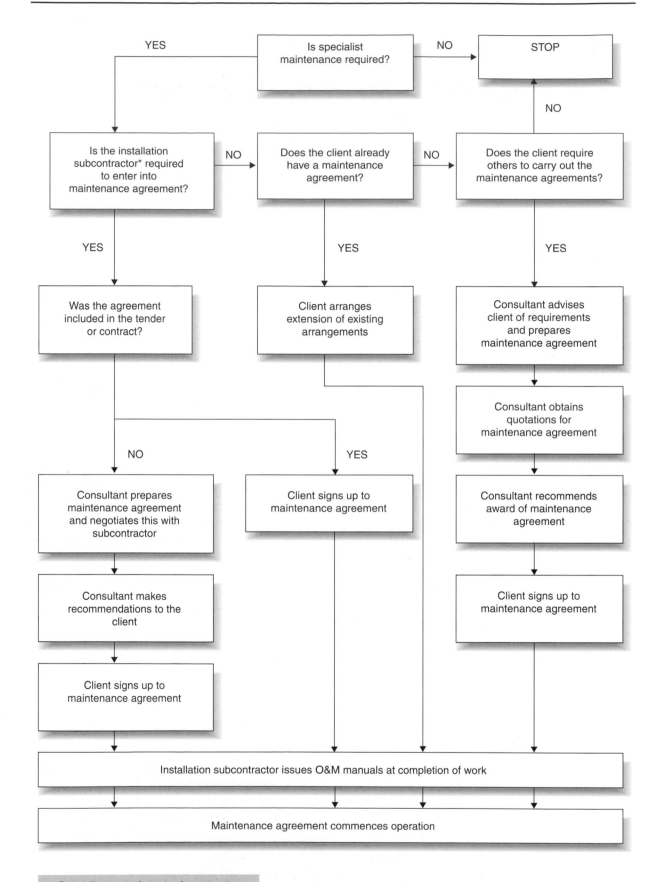

* Specialist, or works or trade contractor

Figure 6.5 Specialist maintenance contracts flow chart

7

Completion, handover and occupation stage

Client's objectives

At this stage the client's aims include agreeing to a handover plan and schedule and client/supplier responsibilities, especially in terms of criteria for acceptance, provision of necessary project documentation, and defects liability, commissioning arrangements, and any instructions as to future occupation. The client is also to agree and implement the handover method and agree a defect rectification plan if necessary and transfer of documentation. Also an initial post-occupancy review may be undertaken at this stage to highlight any immediate issues for rectification.

Completion

Completion and handover are very much interlinked. This is the final stage of work executed by the contractors and consultants prior to acceptance of the facility by the client. They are carried out under the continued co-ordination and supervision of the project manager, in close working relationship with the consultants. The project manager maintains required liaison between and acts on behalf of the parties concerned (e.g. client/user). Occupation organised by the client's occupation co-ordinator is usually preceded by an accommodation schedule of works which can consume anything up to 3% of the construction budget. These works may or may not involve the design consultants and may be managed by the project manager or by the client's accommodation manager.

Project management actions

This stage marks the end of the main construction works, and involves the project manager in a number of activities to terminate the construction contract successfully.

■ Ensuring the contract administrator has inspected the works and, if appropriate, has issued the Certificate of Practical Completion. Attached to the Certificate should be a list of outstanding snags and exclusions together with a statement of the time-scale for their final completion. The project manager needs to ensure completion of these final items does not cause disruption to the client's use of the end product.

■ Issue of the Certificate of Practical Completion marks a transfer of responsibility for the end product from the contractor to the client. The project manager needs to ensure the client is prepared for the insurance and security implications of this change of responsibility.

- A Certificate marking the completion of part of the works can be issued at any time during the project. Sectional completion is used for the early handover of part of the end product, e.g. a computer room.

- Following issue of the Certificate of Practical Completion, the project manager should ensure the final account process for the completed works is concluded with the contractor as rapidly as possible. The final account is a reconciliation of the tendered works and the scope of the works finally instructed, and takes account of variations to the contract issued during the course of the project. While assessment of the cost and time implications of these contractual entitlements is initially made by the contract administrator, in the event of the contractor being unhappy at the proposed settlement, the project manager will be called to arbitrate.

- The final account process involves consideration of claims for additional monies and time made outside of the contract. The project manager, who will make a recommendation to the client for any awards will consider these claims for consequential loss. The project manager has a duty to monitor the legal liability of the client throughout the construction work.

- Ensuring that during the Defects Liability Period there is a system in place for the client to report defects and for the contractor to carry out rectification works. At the end of the defects period, the project manager should ensure the contract administrator carries out a final inspection and, if appropriate, issues the Final Certificate.

- At practical completion a number of significant documents are handed over from the contractor to the client. The project manager needs, on behalf of the client, to ensure firstly, that these documents are available and secondly, that they are to the necessary quality:

  ○ the project's health and safety file

  ○ 'as built' drawings together with all relevant specifications, etc.

  ○ operating and maintenance manual, consisting of details of maintenance schedules, operating instructions, manufacturer's details

  ○ warranties and guarantees from suppliers

  ○ copies of statutory authority approvals and consents

  ○ test and commissioning documentation.

Actions by consultants

Consultants should carry out the following actions:

- Inspect, as appropriate, the work for which they have design responsibility and report to the design team leader, with copy to the project manager, on progress and compliance with contract provisions, highlighting any corrective action to be taken.

- Inspect work at the practical completion stage, produce the outstanding work schedule and sign off, certifying, subject to completion of works listed in the schedule. As a general rule, a certificate of practical completion should not be issued if there are incomplete or defective works outstanding.

- Modern air-conditioned facilities and control systems require a range of external temperatures and full occupation to test their adequacy and stability, i.e. summer and winter working. Certificates of Practical Completion should therefore be qualified and inevitably some final commissioning will take place to services after occupation but before the issue of a Final Certificate.

■ Inspect the work at the end of the contract Defects Liability Period, compile a defects schedule and subsequently confirm that: (1) all defects have been rectified; (2) any omissions have been made good; and (3) all necessary repairs have been carried out.

Planning and scheduling handover

The overall objective is to schedule the required activities to achieve a co-ordinated and satisfactory completion of all work phases within the cost plan. This has to be meshed with the logistical planning of the client's occupation co-ordinator and any accommodation schedule of work to be completed prior to occupation.

Generally, construction projects can be subject to phased (sectional), as well as practical completion. The relevant procedures applied depend on the nature and complexity of the project, and/or requirements of the users. In effect, phased completion means the practical completion for each specific phase of construction. However, this must not:

■ prevent or hinder any party from commencing, continuing or completing their contractual obligations

■ interfere with the effective operation of any plant or services installations.

In cases of phased completion handover, the user/tenant is usually responsible for insuring the works concerned. Upon practical completion handover, the whole of the insurance premium becomes the users' responsibility.

Procedures

The actual practical completion and handover procedures applicable to a specific project will be detailed by the project manager in the handbook for the project concerned (see Appendices 18 of Part 1 and E of Part 2 for typical examples). However, the main aspects of completion and handover will generally cover the minutiae of the following activities:

■ Preparation of lists identifying deficiencies, e.g. unfinished work, frost damage, and materials, goods, and workmanship not in accordance with standards.

■ All remedial and completion work carried out within the specified time under the direct supervision of nominated qualified and experienced personnel.

■ Monitoring and supervising completion and handover against the schedule.

■ The provision of the required number of:

  ○ copies of the health and safety file

  ○ 'as built' and 'installed' record drawings, plans, schedules, specifications, performance data and tests results

  ○ commissioning and test reports, calibration records, operating and maintenance manuals, including related health, safety and emergency procedures

  ○ planned maintenance schedules and specialist manufacturers' working instructions.

■ Monitoring proposals for the training of engineering and other services staff and assistance in the actual implementation of agreed schemes.

■ Ensuring that handover takes place when all statutory inspections and approvals are satisfactorily completed but does not take place if the client/tenant cannot have beneficial use of the facility, i.e. not before specified defects are made good, indicating likely consequences and drawbacks of premature occupation.

■ Setting up procedures to monitor and supervise any post-handover works, which do not form part of the main contract, and to monitor the Defects Liability Period.

■ Initiating, in close co-operation with the relevant consultants, contra-charging measures in cases of difficulties with completing outstanding works or making good any defects.

■ Monitoring progress of final accounts by assisting in any controversial aspects or disputes, and by ascertaining that draft final accounts are available on time and are accurate.

■ Reviewing progress at regular intervals, to facilitate a successful final inspection, and the issuing of a Final Certificate.

■ Establishing the plan for post-completion project evaluation and feedback from the parties to the contract for the post-completion review project close-out report.

Client commissioning and occupation

Having accepted the constructed structure from the contractor at practical completion, the client finally has to prepare the facilities ready for occupation. This stage of the project life-cycle comprises three major groups of tasks: client accommodation works, operational commissioning and migration.

■ In order to allow as much time as possible for the client organisation to develop their detailed requirements, or to reflect their latest business 'shape', it is common for the client to organise a further project to carry out accommodation works. It is likely the project manager will be involved to manage the project team established to carry out these works. Often this team will be separate from the main project team and will comprise personnel with greater experience of operating in a finished project environment.

Typical elements of client accommodation works for an office building would be:

○ fitting out of special areas
 – restaurant/dining areas
 – reception areas
 – training areas
 – executive areas
 – post rooms
 – vending areas

○ installation of IT systems
 – servers
 – desktop PCs
 – telecommunications equipment

- fax machine
- audio-visual and video conferencing
- demountable office partitions
  - furniture
  - specialist equipment
  - security systems
  - artwork and planting.

Operational commissioning

The principles of client commissioning and occupation should be determined at the feasibility and strategy stage. Client commissioning (as with occupation, which usually follows on as a continuous process) is an activity predominantly carried out by the client's personnel, assisted by the consultants as required.

The objective of client commissioning is to ensure that the facility is equipped and operating as planned and to the initial concept of the business plan established for the brief. This entails the formation, under the supervision of the client's occupation co-ordinator, of an operating team early in the project so that requirements can be built into the contract specifications. Ideally, the operating team is formed in time to participate in the design process. (Their role is identified in Part 2, supported by a checklist in Appendix F.)

Main tasks

The main tasks are as follows:

■ Establishing the operating and occupation objectives in time, cost, quality and performance terms. Consideration must be given to the overall implications of phased commissioning and priorities defined for sectional completions, particular areas/services and security.

■ Arranging the appointment of the operating team in liaison with the client. This is done before or during the detailed design stage, so that appropriate commissioning activities can be readily included in the contract.

■ Making sure at budget stage that an appropriate allowance for client's commissioning costs is made. Accommodation schedule of works can consume as much as 3% of the total construction budget.

■ Preparing role and job descriptions (responsibilities, time-scales, outputs) for each member of the operating team. These should be compatible with the construction schedule and any other work demands on members of the operating team.

■ Co-ordinating the preparation of a client's commissioning schedule and an action list in liaison with the client, using a commissioning checklist (see Appendix F, Part 2).

■ Arranging appropriate access, as necessary, for the operating team and other client personnel during construction, by suitable modification of the contract documents.

■ Arranging co-ordination and liaison with the contractors and the consultants to plan and supervise the engineering services commissioning, e.g. preparation of new work practices manuals, staff training and recruitment of additional staff if necessary; the format of all commissioning records; renting equipment to meet short-term demands; overtime requirements to meet the procurement plan; meeting the quality and performance standards, all as defined in Chapter 6.

- Considering early appointment/secondment of a member of the client management team to act as the occupation co-ordinator; this ensures a smooth transition from a construction site to an effectively operated and properly maintained facility (see Appendix 19 for an introduction to facilities management).

- Before the new development can be occupied, the client needs to operationally commission various elements of the development. This involves setting to work various systems and preparing staff ready to run the development and its installations:

  ○ transfer of technology

  ○ checking voice and data installation are operational

  ○ stocking and equipping areas such as the restaurant

  ○ training staff for running various systems

  ○ training staff to run the property.

- Also part of the client's operational commissioning is the obtaining of the necessary statutory approvals needed to occupy the building, such as the Occupation Certificate and the Environmental Health Officer's approval of kitchen areas (if applicable).

- Occupation of developed property is dependent upon detailed planning of the many spaces to be used. For office buildings this space-planning process is developed progressively throughout the project life-cycle.

  Final determination of seating layouts is delayed until the occupation stage in order to accommodate the latent changes to the client's business structure.

  A typical space-planning process consists of:

  ○ confirming the client's space standards including policy on open plan and cellular offices

  ○ confirming the client's furniture standards

  ○ determining departmental headcount and specific requirements

  ○ determining an organisational model of the client's business, reflecting the operational dependencies and affinities

  ○ developing a building stacking in order to fit the gross space of each department within the overall space of the building

  ○ developing departmental layouts to show how each department fits the space allocated to it

  ○ developing furniture seating layouts in order to allocate individual names to desks.

  It is essential that for each of these stages the client organisation in the form of user liaison groups has a direct involvement and approves each stage.

- Moving or combining businesses into new premises is a major operation for a client. During the duration of the moves there is potential for significant disruption to the client's business. The longer the move period, the greater the risks to the client. Migration therefore requires a significant level of planning. Often the client will appoint a manager separate from the new building project to take overall responsibility for the migration. For major or critical migrations, the client should consider the use of specialist migration consultants to support their in-house resource.

■ During the planning of the migration a number of key strategic issues need to be addressed:

  ○ determining how the building will be occupied

  ○ establishing the timing of the moves

  ○ identifying the key activities involved in the migration and assigning responsible managers

  ○ determining move groups and sequence of moves to minimise business disruption

  ○ determining the project structure for managing the move

  ○ identifying potential risks that could impact on the moves

  ○ involving and keeping the client's staff informed.

As some of these strategic issues could have an impact on the timing and sequencing of the main building works, it is important to address them early in the project life-cycle.

■ The final part of occupation is the actual move management. This involves the appointment of a removal contractor, planning the detailed tactics of the moves, and supervision of the moves themselves.

■ The overall period that the moves will be undertaken over is determined by the amount of 'effects' to be transferred with each member of the staff and by the degree of difficulty in transferring IT systems for each move group.

■ A critical decision for the client during the occupation stage is the point at which a freeze is imposed on space planning and no further modifications are accommodated until after migration has been achieved.

It is likely that the factor having most impact on the timing of the freeze date will be the setting up of individual voice and data system profiles.

It is common for clients to impose an embargo on changes on both sides of the migration and for the client then to carry out a post-migration sub-project to introduce all the required changes required by departments.

Client occupation

Occupation should follow a very carefully planned logistical schedule managed by the incoming user of the facility following completion of construction. This can be put under the overarching control of the project manager or can be headed by an appointed occupation co-ordinator.

Unlike many other project management activities, occupation involves employees themselves and is impacted by the style of management and culture of the user's organisation. Consequently, well-executed planning with their involvement in the process can result in better management/employee relations, bringing a greater feeling of participation and commitment to the workforce.

It is normal to produce an operational policy document in the planning stages which is a blueprint for implementation of the business plan. In particular it sets out services which will be kept in-house and those services which will be contracted out and how they will be procured.

The arrangements for occupation and migration from one facility to another will, on many projects, be predetermined by space-planning exercises carried out in the initial design stages by the design team or space-planning consultant. The guidance given in this chapter should be put in the context of the overall planning of a client's needs for a particular facility.

This will follow from:

■ strategic analytical briefing

■ detailed briefing (departmental level)

and lead to criteria such as:

■ quantifying spatial requirements

■ physical characteristics for each department/sector

■ critical affinity groupings

■ extent of amenities

■ workspace standards

■ office automation strategies

■ security/public access

■ furniture, fittings and equipment (FF&E) schedules.

On complex projects this can be taken one stage further to the production of room data sheets which form the basis of the design brief, equipment transfer or purchase, movement of personnel and facilities management.

The procedure outlined below gives a typical approach, which may need to be interpreted in order to harmonise with the practices and expectations of the users. Nevertheless, change in established practices is encouraged where doing so will smooth the process and make it more effective.

Occupation can be divided into four stages as explained below, and shown in Figures 7.1–7.4 on the following pages.

The following services are often outsourced on renewable annual or 3-year contracts:

○ reception and telephony

○ security

○ cleaning

○ building management and operation of services and equipment

○ maintenance

○ IT support

○ catering and waste management

○ landscaping and ground maintenance

○ transport and courier services.

Structure for implementation

Structure for implementation means the appointment of individuals and groups to set out the necessary directions, consultation and budget/cost parameters; Figure 7.1 gives an example.

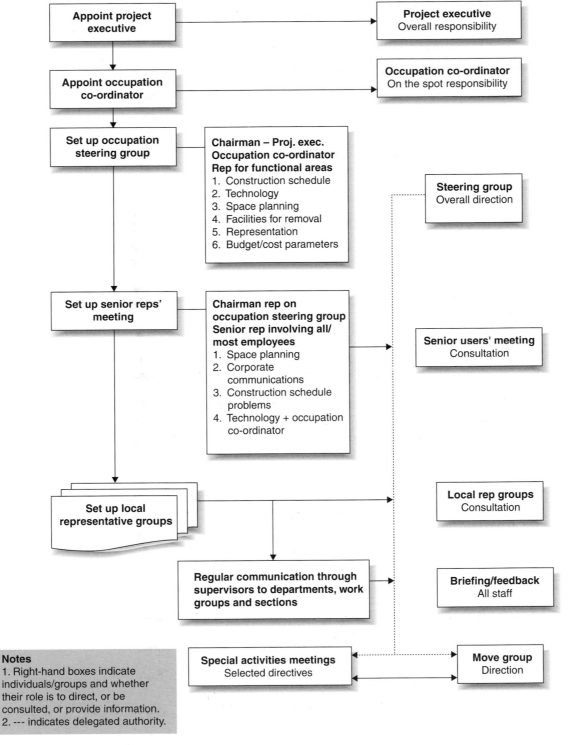

Figure 7.1 Occupation: structure for implementation

Scope and objectives

Scope and objectives means deciding what is to be done, considering the possible constraints and reviewing as necessary; Figure 7.2 gives an example.

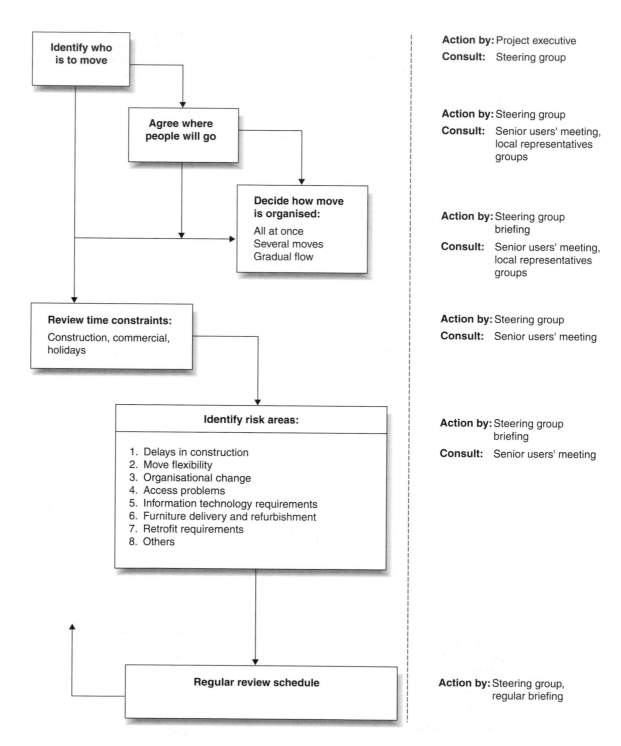

Figure 7.2 Occupation: scope and objectives

Methodology

Methodology is how the whole process will be achieved: identification of individual or groups of special activities and their task lists aimed at defining the parameters and other related matters, e.g. financial implications; Figure 7.3 gives an example.

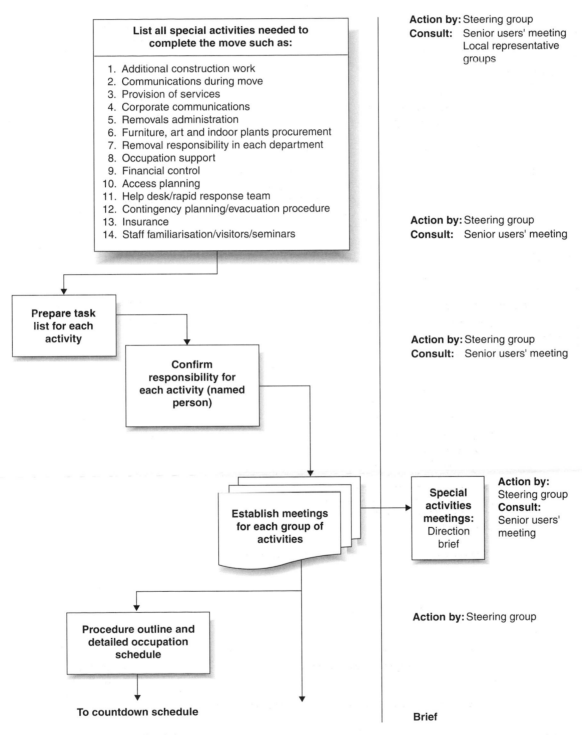

Figure 7.3 Occupation: methodology

Organisation and control

Organisation and control means carrying out the process and keeping schedule and budget/cost under review; Figure 7.4 gives an example.

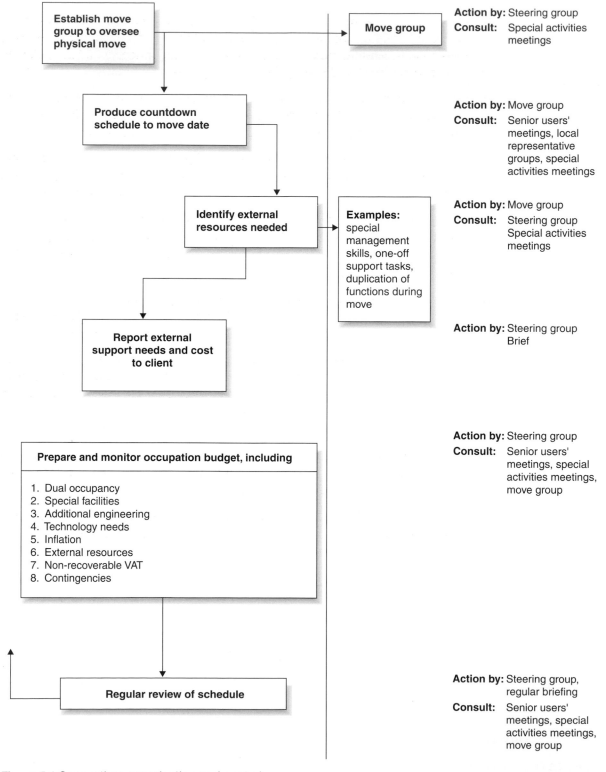

Figure 7.4 Occupation: organisation and control

The individuals and groups likely to be concerned are as follows:

■ *Project executive*: appointed by the client/tenant at the director/senior management level and responsible for the complete process.

■ *Occupation co-ordinator*: project manager appointed, or existing one confirmed by the 'client', with 'on the spot' responsibility.

■ *Occupation steering group*: chaired by the project executive and consisting of occupation co-ordinator and a few selected senior representatives covering the main functional areas. Concerned with all major decisions but subject to any constraints laid down by the client, e.g. financial limits.

■ *Senior representatives' meeting*: chaired by one of the functional representatives on the occupation steering group and made up of a few senior representatives covering the majority of employees and the occupation co-ordinator.

■ *Local representative groups*: chaired by manager/supervisor of own group and concerned with providing views related to a particular location or department. Membership to reflect the specific interest of the group at the location.

■ *Special activities meetings*: meetings for individual or group of special activities as identified in 'methodology'. A single person will be made responsible for achieving all the tasks which make up a special activity and will chair the respective meetings.

■ *Move group*: responsible for the overall direction of the physical move, having been delegated by the occupation steering group, the task of detailed preparation and control of the move schedule including its budget/cost.

■ *Briefing groups*: concerned with effective and regular communication with all employees to provide information to work groups/sections by their own managers or supervisors, so that questions for clarification are encouraged. Special briefings may also be vital, especially during the build-up to occupation.

On many projects, the individuals and groups identified above may be synonymous with those given under client commissioning, e.g. for commissioning team read occupation steering group and vice versa.

Figures 7.1–7.4 provide an at-a-glance summary of the occupation process and Appendix F of Part 2 provides checklists for a typical control system.

8 Post-completion review/ project close-out report stage

Client's objectives

The client's aims at the closing stage of a project should include:

■ to measure performance of all aspects of the project and ensure that the value of the knowledge gained can be carried forward to future projects

■ to undertake an initial assessment of the new facility – so as to establish its fitness for purpose and satisfaction of requirements.

Introduction

The objective of the review is to make a thorough assessment of all elements of the project and to draw out or feed back, for the benefit of the client, the project management practice concerned and other team members, any lessons and conclusions for application to future projects, i.e. what could have been done differently to mutual advantage. This review/report is good practice but should not be regarded as mandatory and may not be required by all clients. It is worth involving the design consultants if only to check that the client is getting maximum use out of the facilities provided and in particular that operational costs are at an optimum level. The typical review is likely to consist of the following elements.

Project audit

■ Brief description of the objective of the project.

■ Summary of any amendments to the original project requirements and their reasons.

■ Brief comment on project form of contract and other contractual/agreement provisions. Were they appropriate?

■ Organisation structure, its effectiveness and adequacy of expertise/skills available.

■ Master schedule – project milestones and key activities highlighting planned versus actual achievements.

■ Unusual developments and difficulties encountered and their solutions.

■ Brief summary of any strengths, weaknesses and lessons learnt, with an overview of how effectively the project was executed with respect to the designated requirements of:

    ○ cost

- o planning and scheduling

- o technical competency

- o quality

- o safety, health and environmental aspects.

■ Was the project brief fulfilled and does the facility meet the client/user needs? What needs tweaking and how could further improvements be made on a value for money basis?

■ Indication of any improvements which could be made in future projects.

Cost and time study

■ Effectiveness of:

- o cost and budgetary controls

- o claims procedures.

■ Authorised and final cost.

■ Planned against actual costs (e.g. S-curves) and analysis of original and final budget.

■ Impact of claims.

■ Maintenance of necessary records to enable the financial close of the project.

■ Identification of time extensions and cost differentials resulting from amendments to original requirements and/or other factors.

■ Brief analysis of original and final schedules, including stipulated and actual completion date; reasons for any variations.

Human resources aspects

■ Communication channels and reporting relationships (bottlenecks and their causes).

■ Industrial relations problems, if any.

■ General assessment and comments on staff welfare, morale and motivation.

Performance study

■ Planning and scheduling activities.

■ Were procedures correct and controls effective?

■ Staff hours summary:

- o breakdown of planned against actual

- o sufficiency of resources to carry out work in an effective manner.

■ Identification of activities performed in a satisfactory manner and those deemed to have been unsatisfactory.

■ Performance rating (confidential) of the consultants and contractors, for future use.

Project feedback

Project feedback necessarily reflects the lessons learnt at various stages of the project, including recommendations to the client for future projects. Ideally feedback should be obtained from all of the participants in the project team at various stages. If necessary, feedback can be obtained at the end of a key decision-making stage (for example, at the completion of each of the seven stages as outlined in this *Code of Practice*).

The project feedback form should include:

- brief description of the project
- outline of the project team
- form of contract and value
- feedback on contract (suitability, administration, incentives, etc.)
- technical design
- construction methodology
- comments on the technical solution chosen
- any technical lessons to be learnt
- form of consultant appointments
- comments on consultant appointments
- project schedule
- comments on project schedule
- cost plan
- comments on cost control
- change management system
- values of changes
- major source(s) of changes/variations
- overall risk management performance
- overall financial performance
- communication issues
- organisational issues
- comments on client's role/decision-making process
- comments on overall project management including any specific issues
- other comments
- close-out report.

It has to be remembered that the purpose of project feedback is not only to express what went wrong and why, but also to observe what has been achieved well, and if (and how) that can be improved in future projects, i.e. continuous improvement.

PART 1 APPENDICES

APPENDIX 1 Typical terms of engagement

Job title: Project manager

Date effective:

General objective

Acting as the client's representative within the contractual terms applicable, to lead, direct, co-ordinate and supervise the project in association with the project team.

The project manager will ensure that the client's brief, all designs, specifications and relevant information are made available to, and are executed as specified with due regard to cost by, the design team, consultants and contractors (i.e. the project team), so that the client's objectives are fully met.

Relationships

Responsible and reporting to

Client

Subordinates

Practice support staff, secretarial/clerical staff

Functional

Fully integrated working with any project support staff who are not line subordinates.

- Liaison, as required/expedient with relevant client's staff, e.g. legal, insurance, taxation.

- Full interdependent co-operation with:

  (a) design team and consultants

  (b) contractors.

External

Liaison with local or other relevant authority on matters concerning the project.

Contact with suppliers of construction materials/equipment, in order to be aware of the most efficient and cost-effective application, and working methods.

Contact with:

(a) Client's information and communication technology (ICT) team or other higher technology sources, able to provide expertise on the application of advanced technology in the design and/or construction processes of the project (e.g. communications, environment, security and fire prevention/protection systems).

(b) And preferably, membership of appropriate professional bodies/societies.

Authority

The definition of the authority of the project manager is a key requirement in enabling him or her to manage the successful achievement of the client's objectives. The extent must be clearly defined. A distinction should be drawn between the responsibility that the project manager may have which concerns his or her accountability for different aspects of the project, and the authority which will determine the ability of the project manager to control, command and determine the commitment of resources to the project. The full extent of the responsibility and authority vested in the project manager will depend on the terms and duties included in the project management agreement.

The extent of the project manager's responsibility and authority may be balanced, but the two may be unequal. Frequently, the project manager may have extensive responsibility in an area that does not carry commensurate authority, or vice versa.

The authority of the project manager should be defined regarding his or her obligations to issue instructions, approve limits of expenditure, and when to notify the client and seek the instructions of the client in matters relating to:

■ the schedule and time taken to complete the project

■ expenditure and costs, including development budget, project cost plan, and financial rewards and viability

■ designs, specifications and quality

■ function

■ contractors' contracts

■ consultants' appointments

■ assignment of contracts or appointments

■ administrative procedures, including issuing or signing of correspondence, certification and other project documentation

The client and the project manager should give careful consideration to the authority that will be necessary to ensure the successful achievement of the client's objectives and, if necessary, establish appropriate lines of authority and communication within the client organisation to facilitate the implementation of agreed procedures.

Detailed responsibilities and duties

1 Analysis of the client's objectives and requirements, assessment of their feasibility and assistance in the completion of project brief and establishment of the capital budget.

2 Formulation, for the client's approval, of the strategic plan for achieving the stated objectives within the budget, including, where applicable, the quality assurance scheme.

3 Generally keeping the client informed, throughout the project, on progress and problems, design/budgeting/construction variations, and such other matters considered to be relevant.

4 Participation in making recommendations to the client, if required, in the following areas:

   (a) The selection of the consultants as well as in the negotiation of their terms and conditions of engagement.

   (a) The appointment of contractors/subcontractors, including the giving of advice on the most suitable forms of tender and contract.

5 Preparation for the client's approval of the following items:

   (a) The overall project schedule embracing site acquisition, relevant investigations, planning, pre-design, design, construction and handover/occupation stages.

   (b) Proposals for architectural and engineering services. The project manager will monitor progress and initiate appropriate action on all submissions concerned with planning approvals and statutory requirements (timely submission, alternative proposals and necessary waivers).

(c)   The project budget and relevant cash flows, giving due consideration to matters likely to affect the viability of the project development.

6   Finalisation of the client's brief and its confirmation to the consultants. Providing them with all existing and, if necessary, any supplementary data on surveys, site investigations, adjoining owners, adverse rights or restrictions and site accessibility/traffic constraints.

7   Recommending to the client and securing approval for any modifications or variations to the agreed brief, approved designs, schedules and/or budgets resulting from discussions and reviews involving the design team and other consultants.

8   Setting up the management and administrative structure for the project and thereby defining:

- responsibilities and duties, as well as lines of reporting, for all parties

- procedures for clear and efficient communication

- systems and procedures for issuing instructions, drawings, certificates, schedules and valuations and the preparation and submission of reports and relevant documentary returns.

9   Agreeing tendering strategy with the consultants concerned.

10   Advising the client as necessary on the following items:

- The progress of the design and the production of required drawings/information and tender documents, stressing at all times the need for a cost-effective approach to optimise costs in construction methods, subsequent maintenance requirements, preparation of tender documents and performance/workmanship warranties.

- The correctness of tender documents.

- The prospective tenderers pre-qualified by the design team and other consultants involved, obtaining additional information if pertinent and confirming accepted tenders to the client and the consultants.

- The preliminary construction schedule for the main contractors, agreeing any revisions to meet fully the client's requirements and releasing this to the project team for action.

- The progress of all elements of the project, especially adherence to the agreed capital and sectional budgets, as well as meeting the set standards and initiating any remedial action.

- The contractual activities the client must undertake, including user study groups and approval/decision points.

11   Establishing with the quantity surveyor the cost monitoring and reporting system and providing feedback to the other consultants and the client on budget status and cash flow.

12   Organising and/or participating in the following activities:

- Presentations to the client, with advice on and securing approval for the detailed design of fabrics, finishes, fitting-out work and the environment of major interior spaces.

Part 1 appendices

    ○   All meetings with the project team and others involved in the project (chairing or acting as secretary) to ensure:

    (a)   an adequate supply of information/data to all concerned

    (b)   that progress is in accordance with the schedule

    (c)   that costs are within the budgets

    (d)   that required standards and specifications are achieved

    (e)   that contractors have adequate resources for the management, supervision and quality control of the project

    (f)   that the relevant members of the project team inspect and supervise construction stages as specified by the contracts.

13   Responsible for:

○   preparation of the project handbook

○   achieving good communications and motivating the project team

○   monitoring progress, costs and quality and initiating action to rectify any deviations

○   setting priorities and effective management of time

○   co-ordinating the project team's activities and output

○   monitoring project resources against planned levels and initiating necessary remedial action

○   preparing and presenting specified reports to the client

○   submitting to the client time sheets and other data on costing and control processes, including required returns and all other relevant information

○   approving, in collaboration with the project team and within the building contract provisions, any sublet work

○   identifying any existing or potential problems, disputes or conflicts and resolving them, with the co-operation of all concerned in the best interests of the client

○   recommending to the client the consultants' interim payment applications and monitoring such applications from contractors

○   monitoring all pre-commissioning checks and progress of any remedial defects liability work and the release of retention monies

○   verifying with the project team members concerned any claims for extensions of time or additional payments and advising the client accordingly

○   checking consultants' final accounts before payment by the client

○   monitoring the preparation of contractors' final accounts, obtaining relevant certificates and submitting them for settlement by the client

○   ensuring the inclusion in the contract and subsequently requesting the design team, consultants and contractors to supply the client with as-built and installed drawings, operating and maintenance manuals, and health and safety file, as well as ensuring arrangements are made for effective training of the client's engineering and maintenance staff, i.e. facilities management

14 Taking all appropriate steps to ensure that site contractors and other regular or casual workers observe all the rules, regulations and practices of safety and fire prevention/protection. Exercising 'good site housekeeping' at all times.

15 Participating in the final cost reconciliation or final account of the project and taking such action as directed or required.

Extra-project activities

Participating in informal discussions with own and other practices, as well as the client's staff, on technical details, methods of operations, problem solving and any other pertinent actions relevant to present or previous projects, in order to exchange views/knowledge conducive to providing a more effective overall performance.

The project manager has responsibility for the following areas:

(a) Personnel matters relating to his staff, including appraisal/reviews, training/development and job coaching and counselling, as defined by the client and/or project management practice guidelines and procedures.

(b) Updating self and staff in new ideas relating to project management, including management/supervisory skills and practice generally, business, financial, legal and economic trends, the latest forms of contract, planning and Building Regulations, as well as advances in construction techniques, plant and equipment.

Terms of engagement – the services contracts

1. RICS Project Management Agreement (3rd edn) (1999)

2. APM Terms of Appointment for a Project Manager (1998)

3. NEC Professional Services Contract (PSC) (1998)

4. RIBA Form of Appointment for Project Managers (1999)

5. NHS Estates Agreement for the Appointment of Project Managers for commissions for construction projects in the NHS

1. RICS Project Management Agreement (3rd edn) (1999)

The Royal Institution of Chartered Surveyors (RICS) was the first to publish the 'Project Management Agreement & Conditions of Engagement' as a model for use by clients and project managers.

2. APM Terms of Appointment for a Project Manager (1998)

This was first issued by the APM in 1998. The terms of engagement are broadly divided into two sections: (i) the core terms and (ii) the schedule of services.

The core terms include the form of agreement, general terms, the schedule of particulars and the fee schedule.

The schedule of services includes four different scopes of services to suit different needs and requirements: model C for construction projects, model I for information technology projects, model M for manufacturing projects and model S for small projects.

3. *New Engineering Contract – Professional Services Contract (1998)*

Another example of the terms of engagement for the project manager is the Professional Services Contract (PSC), which is a part of the Engineering and Construction Contract – New Engineering Contract (NEC – ECC).

4. *RIBA Form of Appointment for Project Managers (1999)*

This was published in 1999 by RIBA as a part of its appointing documentation for construction project management practitioners.

5. *NHS Estates Agreement for the Appointment of Project Managers for commissions for construction projects in the NHS*

The provision of project management services as described in the NHS Estates' Agreement for the appointment of project managers includes the overall planning, control and co-ordination of a project from inception to completion with the object of meeting the client's requirements and ensuring completion on time, within cost and to the required quality standards.

APPENDIX 2   Health and safety in construction including **CDM** guidance

Generally, the laws, which govern health and safety relate to all construction activities (including design), and are not industry specific. There are several Acts and Regulations involved and information to access them is provided at the end of this briefing sheet.

Some of the principal Acts which deal with health, safety and welfare in construction are:

■ Health and Safety at Work Act 1974

■ Mines and Quarries Act 1954

■ Factories Act 1961

■ Offices, Shops and Railways Premises Act 1963

■ Employers Liability Acts – various

■ Control of Pollution Act 1974

■ Highway Act 1980

■ New Roads and Streetworks Act 1991.

The fundamental Act governing health and safety in construction is the **Health and Safety at Work Act 1974**. This Act has some 62 separate Regulations and it is not possible to deal with such a large subject area here, however, the principal regulations of this act, which affect design and construction, are:

■ Management of Health and Safety at Work Regulations 1992

■ Construction (Design and Management) Regulations 1994 [known as CDM Regulations]

■ Construction (Health, Safety and Welfare) Regulations 1996

Some other related regulations and guides are:

■ Health and Safety Regulations – A Short Guide

■ Reporting of Injuries, Diseases and Dangerous Occurrences Regulations (RIDDOR) 1995

■ The Control of Major Accident Hazards Regulations 1999 (COMAH)

■ The Chemicals (Hazard Information and Packaging for Supply) Regulations 2000 (CHIP)

■ The Health and Safety (Display Screen Equipment) Regulations 1992 – Working with VDUs

■ A guide to the Construction (Health, Safety and Welfare) Regulations 1996 Introducing the Noise at Work Regulations

■ COSHH (Control of Substances Hazardous to Health): a brief guide to the regulations 1994

■ Simple guide to the Provision and Use of Work Equipment Regulations (PUWER 98)

■ Simple guide to the Lifting Operations and Lifting Equipment Regulation (LOLER 98) Workplace Health Safety and Welfare

LIVERPOOL JOHN MOORES UNIVERSITY
LEARNING & INFORMATION SERVICES

■ A short guide to the Personal Protective Equipment at Work Regulations 1992

■ Signposts to the Health and Safety (Safety, Signs and Signals) Regulations 1996

■ Approved Code of Practice (ACOP) on various work-related areas as published and amended by the Health and Safety Executive (HSE) and approved by the Health and Safety Commission (HSC).

CDM Regulations

These regulations became law in England and Wales in April 1995. They control how building projects are run in order to protect the health and safety of construction workers and future building users. The regulations place strict duties on clients, designers and others. This document is a *brief summary* of the regulations where they affect you: **the client**.

When do the regulations apply?

The regulations apply to all building work when you have reason to believe that either of the following will occur:

1. the project will last more than 30 days, or will involve any demolition or dismantling of a structure;

2. there will be five or more persons engaged on-site at any one time;

The regulations do not apply if you are having your own house built or altered (unless there is some demolition) although you must still notify the HSE about your project.

CDM applies to **all** demolition and structural dismantling work, except where it is undertaken for a domestic client.

CDM applies to most construction projects. There are a number of situations where CDM does not apply. These include: some small-scale projects which are exempt from some aspects of CDM; construction work for domestic clients (although there are always duties on the designer, and the contractor should notify HSE where appropriate); construction work carried out inside offices and shops, or similar premises, that does not interrupt the normal activities in the premises and is not separated from those activities; the maintenance or removal of insulation on pipes, boilers or other parts of heating or water systems.

Can I appoint someone else to carry out my duties?

You can appoint an agent to act on your behalf as client if you wish. If you do, you should ensure that they are competent to carry out your duties. If you appoint an agent, they should send a written declaration to HSE.

It should:

■ state that the agent is acting on your behalf

■ give the name and address of the agent

■ give the exact address of the construction site

■ be signed by or on behalf of your agent.

Can I appoint myself to carry out other duties?

You can appoint yourself as planning supervisor and/or principal contractor providing you are competent and adequately resourced to comply with their health and safety responsibilities. You need to be a contractor in order to appoint yourself as principal contractor.

Table 1 Summary of the key duties of the client, principal contractor, planning supervisor and designer

---

Client

■ To select and appoint a competent planning supervisor and principal contractor.

■ To be satisfied that the planning supervisor and principal contractor are competent and will allocate adequate resources for health and safety.

■ To be satisfied that designers and contractors are also competent and will allocate adequate resources when making arrangements for them to work on the project.

■ To provide the planning supervisor with information relevant to health and safety on the project.

■ To ensure construction work does not start until the principal contractor has prepared a satisfactory health and safety plan.

■ To ensure the health and safety file is available for inspection after the project is completed.

Principal contractor

■ To develop and implement the health and safety plan.

■ To arrange for competent and adequately resourced contractors to carry out the work where it is subcontracted.

■ To ensure the co-ordination and co-operation of contractors.

■ To obtain from contractors the main findings of their risk assessments and details of how they intend to carry out high risk operations.

■ To ensure that contractors have information about risks on site.

■ To ensure that the workers on site have been given adequate training.

■ To ensure that contractors and workers comply with any site rules which may have been set out in the health and safety plan.

■ To monitor health and safety performance.

■ To ensure that all workers are properly informed and consulted.

■ To make sure only authorised people are allowed onto the site.

■ To display the notification of the project to HSE.

■ To pass information to the planning supervisor for the health and safety file.

Planning supervisor

■ To co-ordinate the health and safety aspects of project design and the initial planning.

■ To ensure that a health and safety plan for the project is prepared before a principal contractor is employed.

■ To take reasonable steps to ensure co-operation between designers, for the purposes of health and safety.

■ To assist with enquiries concerning the principal contractor and designer and to advise on the health and safety plan before the construction phase commences.

■ To notify the project to the HSE.

■ To ensure that the health and safety file is prepared and delivered to the client at the end of the project.   *(cont'd overleaf)*

Table 1 Summary of the key duties of the client, principal contractor, planning supervisor and designer (*cont'd*)

Designer
■ To alert clients to their duties.
■ To consider during the development of designs the hazards and risks which may arise to those constructing and maintaining the structure.
■ To design to avoid risks to health and safety so far as is reasonably practicable.
■ To reduce risks at source if avoidance is not possible.
■ To consider measures which will protect all workers if neither avoidance nor reduction to a safe level is possible.
■ To ensure that the design includes adequate information on health and safety.
■ To pass this information on to the planning supervisor so that it can be included in the health and safety plan and ensure that it is given on drawings or in specifications, etc.
■ To co-operate with the planning supervisor and where necessary other designers involved in the project.

Table 2 Suggested enquiries to be made by the client in the appointment of relevant personnel

Planning supervisor
■ His/her knowledge of construction practice relevant to the project.
■ His/her familiarity and knowledge of design function.
■ His/her knowledge of health and safety issues, particularly in preparing health and safety plans.
■ His/her ability to work with and co-ordinate activities of different designers.
■ The number, experience and qualifications of people to be employed, internally and from other sources.
■ The management systems to be used.
■ The time to be allowed to carry out the different duties.
■ The technical facilities available to aid staff in carrying out duties.
Principal contractor
■ The personnel to carry out or manage the work, their skills, knowledge, experience and training.
■ The time to be allocated to complete various stages of the construction work without risks to health and safety.
■ The manner in which people are to be employed to ensure compliance with health and safety law.
■ The technical and managerial approach for dealing with the risks specified in the health and safety legislation.
Designer
■ His/her knowledge, ability and resources.
■ His/her familiarity with construction processes and the impact of design on health and safety.
■ His/her awareness of relevant health and safety legislation and appropriate risk assessment methods.
■ His/her health and safety policies.
■ The personnel to be employed to carry out the work, their skills and training (this may include external resources where necessary) and review of the design.
■ The time to be allocated to fulfil the various elements of the designer's work.
■ The technical facilities available.
■ The method of communicating design decisions and information on any remaining risks.

Part 1 appendices

Table 3 Considerations and suggested amendments concerning construction documentation

---

Planning supervisor

In appointment:

■ Written confirmation that the planning supervisor

  ○ understands the role and is prepared to perform it

  ○ is aware of and will comply with CDM.

■ A warranty that the planning supervisor has and will maintain the necessary competence and resources.

■ Certain specific services to be itemised separately, e.g. advice on whether health and safety plan has been prepared in accordance with CDM.

■ Consider fee structure.

■ Consider rights of termination (e.g. if principal contractor can take over role of planning supervisor).

■ Make duties clear.

■ Consider if reasonable skill and care is sufficient.

■ Consider copyright; it is vital that the client has the ability to obtain rapid delivery of the health and safety plan and the health and safety file.

■ Check that professional indemnity insurance extends to role of planning supervisor.

■ Consider whether collateral warranties are needed for various interested parties, e.g. funders, purchasers and/or tenants, and if so the form of such warranties.

Principal contractor

■ Health and safety plan should be sufficiently developed to form part of tender documentation or similar proposals.

■ Consider whether standard building contracts should be amended specifically to include principal contractor's acknowledgement that he is aware of and will comply with CDM.

■ Certain specific duties to be spelt out in contract, e.g. maintenance of health and safety plan.

■ Where a number of contractors occupy a site either one contractor must be given over-arching control of the site and occupy the role of 'principal contractor' or distinct physical boundaries delineated by hoardings, fencing or scaffolding must be established with each acting as 'principal contractor' within its defined site. The project manager should arrange co-ordination meetings between all the parties on a regular basis in such circumstances.

Designer

In appointment:

■ Consider whether obligation should be included that designers are aware of and will comply with CDM.

■ Consider services to be provided, e.g. co-operation with the planning supervisor and other designers.

■ Consider whether collateral warranties normally given by designers need to be altered to include express reference to compliance with CDM.

Key CDM activities for the client at various project stages

Feasibility, strategy and pre-construction stage

■ Determine whether the project falls within the scope of the Regulations.

■ Appoint a designer and planning supervisor – ensure competency of personnel and that each will allocate adequate resources to health and safety. Appointments should be made as early as practicable.

■ Provide planning supervisor and designers with information relevant to the health and safety of the project, e.g. risk assessment records, health and safety policy.

During these stages the planning supervisor will need to ensure that the project is notified to the HSE (if relevant), that the designers appointed by the client are competent, that the health and safety file is opened and that the pre-tender health and safety plan is prepared before arrangements are made to appoint the principal contractor. The designer will co-operate with the planning supervisor on these issues.

Selection of principal contractor

■ Ensure that the principal contractor is competent and has allowed sufficient resources for health and safety matters by examining pre-qualification criteria in the tender submission. The client should seek the advice of the planning supervisor on this matter.

Construction stage

■ Ensure that construction work does not begin until the principal contractor has prepared a satisfactory health and safety plan which complies with the Regulations. The client should seek the advice of the planning supervisor on this matter.

■ Ensure that any activities which may affect the construction work comply with health and safety legislation.

Completion and occupation stage

■ Ensure that the health and safety file is available for inspection. The planning supervisor is responsible for ensuring that the file is prepared, completed and delivered to the client. This should be as soon as practical after substantial completion of the work. A draft copy should be available at practical completion.

Suggested contents for the construction health and safety plan

■ Introduction: description; health and safety objectives; restrictions.

■ Management: who has responsibility for what?

■ Communications.

■ Risks: signifies hazards; risk assessments; method statements; COSHH assessments; permits to work.

■ Other matters: emergencies; welfare; RIDDOR; training; consultation; site rules.

■ Monitoring: Who? When? Reporting? Follow-up? Preventive action.

Suggested contents for the health and safety file

■ 'As built' drawings; diagrams.

Part 1 appendices

APPENDIX 4 Site investigation

This flow chart is used for each of the 10 activities identified in the table below.

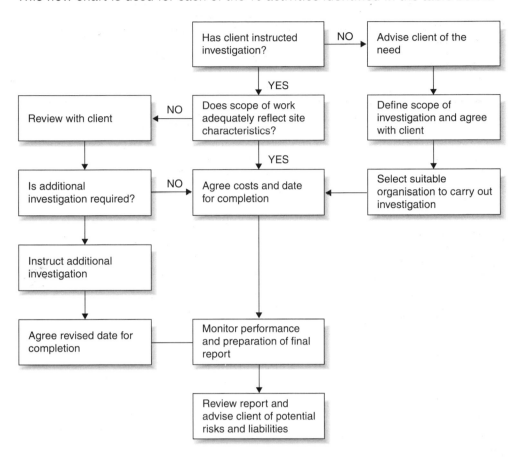

Activities associated with site investigation

Activity	Action by
Site surveys	Land surveyor and structural engineer
Geotechnical investigation	Ground investigation specialist
Drainage and utilities survey	Civil engineering consultant
Contamination survey	Environmental and/or soil specialist
Traffic study	Transportation consultant
Adjacent property survey	Buildings/party walls/rights of light surveyors
Archaeological survey	Local museum or British Museum and other relevant source
Environmental issues	Specialist consultant
Legal aspects	Solicitor
Outline planning permission	Architect

Confirmation that the activities have been successfully completed is the responsibility of the project manager.

Each task can be broken down into a number of specific elements:

Site surveys
- location
- Ordnance Survey reference

- ground levels/contours
- physical features (e.g. roads, railways, rivers, ditches, trees, pylons, buildings, old foundations, erosion)
- existing boundaries
- adjacent properties
- site access
- structural survey
- previous use of site.

Geotechnical investigation
- trial pits
- boreholes and borehole logs
- geology of site including underground workings
- laboratory soil tests
- site tests
- groundwater observation and pumping tests
- geophysical survey.

Drainage and utilities survey
- existing site drainage (open ditch, culvert or piped system)
- extent of existing utilities on or nearest to the site (water, gas, electricity, telecoms)
- extent of any other services that may cross the site (e.g. telephone/data lines, oil/fuel pipelines).

Contamination survey
- asbestos
- methane
- toxic waste
- chemical tests
- radioactive substances.

Traffic survey
- examination of traffic records from local authority
- traffic counts
- traffic patterns
- computer simulation of existing traffic flows
- delay analysis
- noise levels.

Adjacent property survey
- right of light
- party-wall agreements

- schedule of conditions
- foundations
- drainage
- access
- public utilities serving the property
- noise levels (e.g. airports, motorways, air-conditioning equipment).

Archaeological survey
- examination of records
- archaeological remains.

Environmental issues
- effects of proposed development on local environment; environmental impact assessment, where appropriate.

Legal aspects
- ownership of site
- restrictive covenants
- easements, e.g. rights of way, rights of light
- way-leaves
- boundaries
- party-wall agreements
- highways agreements
- local authority agreements
- air rights.

Outline planning permission
- effect of local area plan.

APPENDIX 5 Guidance on EU procurement directives

At EU level the general public procurement rules originate from the EC Treaty, where article 12 prohibits discrimination on the grounds of nationality. Articles 28–30 prohibit quantitative restrictions and measures having an equivalent effect while articles 43–55 ensure the right of free establishment and free exchange of services. These general rules have been complemented by a set of directives, which set out the procedural aspects and are provisioned to be updated from time to time.

Public procurement directives

The public procurement directives are the works directive, the services directive, the supplies directive, the utilities directive and the remedies directives.

1. Council Directive 93/36/EEC of 14th June 1993 co-ordinating the procurement procedures for the award of public supply contracts

2. Council Directive 93/38/EEC of 14th June 1993 concerning the co-ordination of procedures for the award of the public works contracts

3. Council Directive 92/50/EEC of 18th June 1992 relating to the co-ordination of the procedures for the awards of the service contracts

4. Council Directive 89/665/EEC of 21st December 1989 on the co-ordination of the laws, regulations and administrative provisions relating to the application of review procedures to the award of public supply and public works contracts

5. Council Directive 92/13/EEC of 25th February 1992 co-ordinating the laws, regulations and administrative provisions relating to the application of community rules on the procurement procedures of entities operating in the water, energy, transport and telecommunication sectors.

Work is being undertaken to revise the policy directives particularly by bringing together the existing separate directives governing the public-sector procurement into a single consolidated directive, simplifying the rules to make it more easily understood, and adding flexibility – for instance, procurement through competitive dialogue and framework agreements.

Procurement options

The public authorities can use five different kinds of procurement procedure as per the existing set of directives:

■ Open procedure – where all interested suppliers may submit tenders.

■ Restricted procedure – where only invited suppliers may submit tenders.

■ Accelerated procedure – which is somewhat similar to the restricted procedure, but with much shorter deadlines (3–4 weeks instead of 11) and which can only be used in case of special needs for speedy acquisition.

■ Negotiated procedure – where the conditions of the agreements are negotiated with one or several potential suppliers and where the procedure can be used both with and without prior publication; this procedure can only be used under special circumstances as it gravely limits competition.

■ Qualification procedure – which is used under the utilities directive and where suppliers can become pre-qualified if they fulfil certain objective criteria.

The public authorities must use either the lowest price or the most economically advantageous tender (where tender document specifies what attributes apart from price, for instance quality and service, will be given priority) as the award criteria.

OGC *guidance on examples of cases where competition is not needed*

A guidance note on EU procurement policies published by the OGC provides the following examples where the EU competition laws do not take precedence:

■ Works contracts – cases where for technical or artistic reasons or for reasons connected with the protection of exclusive rights the contract can only be carried out by a particular person.

■ Public works concession contracts – requirements are required to be advertised in the *Official Journal*, expressions of interest cannot be rejected on grounds of nationality, but there is no requirement for a competition.

■ Service contracts – competition is not required for certain services and, in addition to the artistic reasons/exclusive rights exemption, competition is not required where the rules of a design contest require a service contract to be awarded to a successful contestant or contestants. There is also greater scope for public authorities to use the competitive negotiated procedure than for works or supplies.

■ Service concession contracts – these fall outside the scope of the EC procurement rules and supply contracts – competition is not required where the artistic reasons/exclusive rights exemption applies. In addition goods manufactured for certain research, experiment, study or development purposes do not have to be procured competitively.

APPENDIX 6 Performance management plan (PMP)

Performance management should be an integrated part of development projects from definition through to monitoring and review.

Where it is not possible to establish direct 'cause and effect' linkages or precise measures of performance, interim measures such as key performance indicators (KPIs) are often used. These could be trends over time, value to the customer, awareness of product or service.

Objectives

The purpose of a PMP is to set out the principles and targets for a schedule against which it delivers its outputs, outcomes and benefits. The plan also defines how the performance criteria will be measured and plans for any divergence management. The plan contains details of the performance management process, performance measurements and the performance information required to establish and monitor delivery.

Performance management process

The performance management process outlines the activities to set direction, which uses performance information to manage better, demonstrates what has been accomplished and sets actions to improve. Performance metrics may be defined using the SMART test (Specific, Measurable, Attainable, Relevant and Timely).

Performance measurements should indicate milestones for measuring progress against goals (maybe at the key decision stages), against target levels of intended accomplishment (target objectives) and against third parties. Measures may need to change as progress is made. Measurement criteria may be defined using the FABRIC test (Focused, Appropriate, Balanced, Robust, Integrated and Cost-effective).

Performance information includes the data, their characteristics, quality, sources and contribution to a measure.

Checklist for PMP

Quality criteria for a performance management plan include:

- Are the objectives, outputs, outcomes or benefits against which to set and monitor performance or achievement of targets clearly defined?

- Can the performance measures be assessed against key objectives?

- Are the performance measures clearly defined, together with target values?

- Is the approach for managing performance complete and does it contain all key elements as a cycle of activities?

- Do the measures and metrics criteria meet pre-defined tests of SMART and FABRIC? (if applicable)

- Is the periodicity of measurement clearly defined?

- Have all standards or techniques to be used for measurement been defined?

- Are the sources of performance information of adequate quality?

Part 1 appendices

■ Is the proposed performance information reliable and/or independently validated?

■ Is the approach to investigation and corrective action to improve unsatisfactory performance clearly defined?

■ Is there an outline of the management organisation and process?

■ Are the resources to collect and analyse performance information clearly defined?

■ Are the roles and responsibilities clearly defined?

Suggested contents for the PMP

The key elements of the PMP will describe a cycle of activities and their outputs:

■ Strategy – defining the aims and objectives of the organisation.

■ Selection of performance measures – identifying the measures which support the quantification of activities over time.

■ Selection of targets – quantifying the objectives set by management, to be attained at a future date.

■ Delivery of performance information – providing a good picture of whether an organisation is achieving its objectives.

■ Reporting information – providing the basis for internal management monitoring and decision making, and the means by which external accountability is achieved.

■ Action to improve – taking action to put things right; feeding back achievements into the overall strategy of the organisation.

APPENDIX 7 Implications of the Housing Grants, Construction and Regeneration Act 1996

The Housing Grants, Construction and Regeneration Act 1996 is applicable to all construction contracts entered into after 1 May 1998. The reforms introduced by this Act can be attributed mainly to the initiative arising after the publication of the Latham Report.

The Act is applicable to all construction operations including site clearance, labour only, demolition, repair works and landscaping.

However, off-site manufacturing, supply and repair or plants in process industries, domestic construction contracts, contracts not in writing and certain other activities including PFI contracts (but not the construction contracts entered into as a result of PFI) have been excluded from the purview of this Act.

The two key areas affected by this Act are the payment procedures and the adjudication in case of disputes.

Payment under the Act

The Act requires that every construction contract must contain the following elements:

- Payment by instalments.

- Adequate mechanism to determine what amounts of payments are due and when.

- Prior notice of amounts due and make up.

- Prior notification (7 days) of intention to withhold payment (set off), giving grounds and amounts.

- Suspension of work (not less that 7 days' notice) for non-payment of payments due.

- All 'pay when paid' clauses outlawed except in the instance where a third party upon whom the payment depends becomes insolvent.

- In the absence of minimum requirements as specified by this Act, the government scheme comes into operation as a default clause.

The government scheme for payment provides for

- Monthly interim payment.

- Due date for interim payments 7 days after the end of the relevant monthly period or making the claim, whichever is later.

- Final due date for interim payment 17 days from due date.

- Notice of amount due not later than 5 days after due date.

- Notice of intention to withhold payment not later than 7 days before final payment date.

Adjudication under the Act

The Act makes it a statutory right to refer any dispute for adjudication. It stipulates that all contracts must contain an adjudication procedure which complies with the Act.

■ Either party can give notice of adjudication at any time regarding any dispute or difference arising under the contract.

■ The contract must provide a timetable for appointment of an adjudicator and referral of dispute within 7 days of the initial notice.

■ The adjudicator must reach a decision within 28 days of referral (up to 42 days if the referring party agrees).

■ The period can be extended only if the parties agree, or at the adjudicator's instigation with the consent of the referring party.

■ The adjudicator is enabled to take necessary initiatives in ascertaining facts and law.

■ The decision of the adjudicator is binding until the dispute is finally determined by legal proceedings or arbitration or by agreement.

■ The parties have the option to agree to accept the adjudicator's decision as final.

■ The government scheme comes into operation as a default mechanism if the minimum requirements as stipulated by the Act are not met.

The government scheme for adjudication provides for

■ Written notice of adjudication.

■ Nature and description of dispute and parties involved.

■ Details of where and when dispute has arisen.

■ Nature of redress sought.

■ Names and addresses of parties to contract.

■ Appointment of adjudicator within 7 days of notice.

■ The same 7 days in which to submit full documentation (referral notice).

■ Oral evidence limited to one representative (may or may not be a lawyer).

■ Adjudicator's decision within 28 days from referral notice or 42 days with the referring party's permission.

■ Parties equally responsible for payment of the adjudicator's fees (unless otherwise determined by the adjudicator).

■ Reasons for the adjudicator's award have to be provided if requested.

■ The decision by the adjudicator is binding pending any final determination by legal proceedings or arbitration, or by mutual agreement in settlement.

■ Parties have to comply with the adjudicator's decision immediately.

A quick look at adjudication

Adjudication is intended to reduce time and costs. Therefore the parties must be prepared for a degree of 'rough justice'. Adjudicators have very wide powers. They can use their initiative and can request further documents from any party, meet and question them, visit the site, appoint experts to help them if necessary (e.g. technical assessors, legal advisers), and issue directions and time-scales. They can adjudicate, with the consent of all parties, on 'related disputes' under different contracts. They can award interest payments.

What can be referred to the adjudicator?

Virtually all kinds of dispute or difference may be referred to an adjudicator, provided they arise 'under the contract'.

These would include: failure to issue notice of sums due or notice of withholding payment, value of interim payments, value of variations, extension of time, loss and expense, set off and contra charges, workmanship, whether or not an instruction is reasonable, etc.

Who pays for the cost of adjudication?

The initial view was that each party must bear his own costs in submitting and presenting his case. However, a recent court case supported a 'costs paid by loser' approach.

How is the adjudicator's decision enforced?

The adjudicator's decision is intended to be binding pending final determination by legal proceedings or arbitration, or by mutual agreement in settlement. Several court cases have shown that the courts intend to support both the Act and the adjudicator by enforcing awards.

It may well be that the mere presence of an adjudication resource will concentrate the minds of those on either side whose stance is less than reasonable, and so enable the parties to go forward with providing the client's end product – the completed project – on time and free of major disputes!

APPENDIX 8   Partnering

Successive review reports on the UK construction industry (particularly the Latham Report, 1994; the Egan Report, 1998; NAO Report, 2001) emphasised the importance of incorporating and initiating partnering arrangements among the supply chain, and being part of the project management process.

What is partnering?

Partnering is defined as a management technique embracing a range of practices designed to promote more co-operative working between contracting parties (Modernising Construction, National Audit Office).

Partnering facilitates team working across contractual boundaries. Its fundamental components are formalised mutual objectives, agreed problem resolution methods, and identifying and incorporating continuous measurable improvements. Developing and maintaining a strong team is essential for the success of partnering.

Project partnering involves the main contractor and the client organisation working together on a single project, usually after the contract has been awarded.

Strategic partnering involves the main contractor and the client organisation working together on a series of construction projects, to promote continuous improvement.

■   Partnering is not contradictory to competition – partners are usually appointed through competition, adopting any of the procurement routes mentioned before. A partnering contract (for example, NEC partnering agreement or a multi-party project partnering contract using the CIC model) can be used to formalise the partnering arrangement.

■   Partnering can promote better value for money by encouraging clients and contractors to work together, minimising the risk of disputes by avoiding an adversarial relationship.

■   Partnering, to be successful, requires all parties to be committed to making the relationship work.

■   Continuous and reliable monitoring should take place to ensure that the relationship is achieving its objectives and those of the project.

Initiating partnering

Partnering is likely to be most appropriate in the following circumstances:

■   On complex projects where user requirements are difficult to specify, for example, where the end-user requirements cannot be clearly defined at the start of the project.

■   For organisations wanting similar facilities repeated over time giving scope for continuous improvements in cost and quality, for example, a chain of stores or warehouses.

■   For projects where construction conditions are uncertain, solutions are difficult to foresee and joint problem solving is necessary, for example where the land is badly contaminated.

■   For individual projects or series of projects where there are known opportunities to drive out waste and inefficiency from the construction process.

Potential benefits

- Greater client and end-user satisfaction.

- Elimination of barriers among the involved parties.

- Replacement of a blame culture with a trusting environment.

- Reduction in waste, thus better value for money.

- Sharing knowledge.

The partnering process

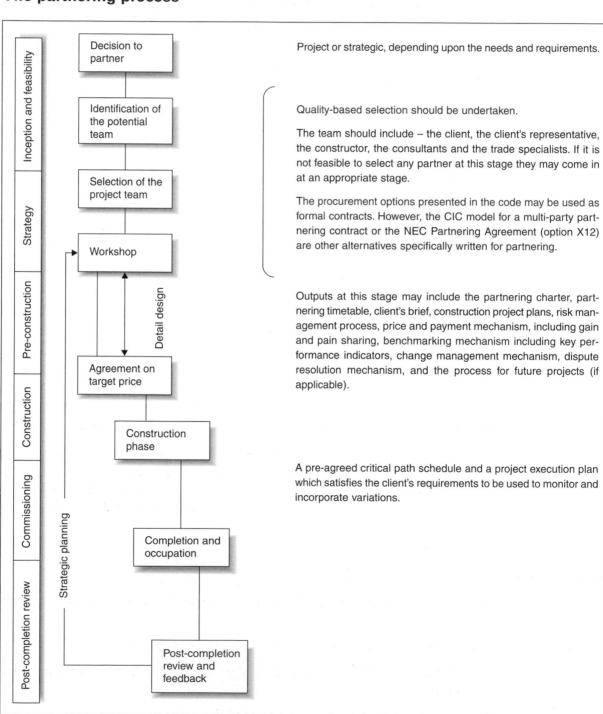

Project or strategic, depending upon the needs and requirements.

Quality-based selection should be undertaken.

The team should include – the client, the client's representative, the constructor, the consultants and the trade specialists. If it is not feasible to select any partner at this stage they may come in at an appropriate stage.

The procurement options presented in the code may be used as formal contracts. However, the CIC model for a multi-party partnering contract or the NEC Partnering Agreement (option X12) are other alternatives specifically written for partnering.

Outputs at this stage may include the partnering charter, partnering timetable, client's brief, construction project plans, risk management process, price and payment mechanism, including gain and pain sharing, benchmarking mechanism including key performance indicators, change management mechanism, dispute resolution mechanism, and the process for future projects (if applicable).

A pre-agreed critical path schedule and a project execution plan which satisfies the client's requirements to be used to monitor and incorporate variations.

APPENDIX 9 Project risk assessment

Risk is inherent in almost all construction projects. Depending on the nature and potential consequences of the uncertainties, measures are taken to tackle them in various ways.

Risk management is a systematic approach to identifying, analysing, and controlling areas or events with a potential for causing unwanted change. It is through risk management that risks to the schedule and/or project are assessed and systematically managed to reduce risk to an acceptable level.

A process of risk assessment and management has to be implemented at an early enough stage to impact on decision making during the development of the project.

Small workshops should put together a list of events which could occur, that threaten the assumptions of the project. A typical list requiring different workshop forums might be:

- revenue
- planning consent
- schedule or time
- design
- procurement and construction
- maintenance and operation.

Assessments should be made on:

- probability of occurrence (%)
- impact on cost–time–function (£ – weeks – other)
- mitigation measures
- person responsible for managing risk
- delete point where risk will have passed (date)
- date for action by – transference – insurance – mitigation (date).

The mitigation measures need evaluating in terms of value for money. Transference should be to the party most capable of controlling it since they will commercially price it lowest. Lateral thinking might shrink or in rare cases eliminate risk. The risk register should be regularly reviewed and be part of all onward decision-making processes.

Risk register

The formal record for risk identification, assessment and control actions is the risk register.

The risk register may be divided into three parts as follows:

- Generic risks: risks that are inherent irrespective of the type or nature of the project.
- Specific risks: risks that are related to the particular project, perhaps identified through a risk workshop involving the project team.

■ Residual risks: this is a list of risks identified above which cannot be excluded or avoided and contingency has to be provided for their mitigation.

The financial effect of such residual risks must be evaluated to determine an appropriate contingency allowance. A time contingency should also be considered.

When using the risk registers, values of occurrence and consequence are assessed using high, medium and low (H/M/L) values. Other forms of assessment include very high, high, medium, low and very low scoring or using numerical values (for example, a ranking of 1–10, with 10 signifying a very high probability/ impact risk and 1 identifying a risk of almost negligible impact/probability).

An example of a part of a risk register is shown below.

Risk Number	Description	Probability of Occurrence %	Unmitigated Impact			Mitigations Adapted	Person Responsible	Last Updated	Delete Point	Date to Action
			Impact Cost	Time	Function					
1										
1.1										
1.2										
1.3										
2										
2.1										
2.2										
2.3										
2.4										
2.5										
2.6										
2.7										
2.8										
2.9										
2.10										
2.11										
2.12										
2.13										
2.14										
2.15										
2.16										
2.17										
2.18										

Often risks are external, arising from events one cannot influence. This particularly applies to markets and revenue projections. Some might even be 'show stoppers'. Value for money studies might throw up options of hedging or taking out special insurance. Obviously the risks of high probability and high impact are the ones on which to concentrate.

An example of a format of a risk mitigation table is shown opposite.

Mitigation action plan

RISK NAME:	DATE:	ISSUE No.:	ISSUED BY:

Risk category/Ref.:		Risk Ownership:	

Risk Evaluation		Probability	Cost	Time	(Other area e.g. environment/ H&S)	Total Score
		Current				
	Projected					

Risk Description:

Risk Mitigation Plan:	By Whom:	Review Point/Milestones:

End date/Time scale

Comments

Project risk assessment checklist

Project _____   Date _____

Overall risk assessment is:

Signatory	Signature	Date
Client		
Project manager		

☐ Normal risk

☐ High risk

Risk consideration	Criteria		Risk assessment		Proposed management of high risk
			Normal	**High**	
1 Project environment					
User organisation	○	Stable/competent	☐ ☐	☐ ☐	
	○	Poor/unmotivated/untrained			
User management	○	Works as a team	☐ ☐	☐ ☐	
	○	Factions and conflicts			
Joint venture	○	Client's sole contractor	☐ ☐	☐ ☐	
	○	Third party involved			
Public visibility	○	Little or none	☐ ☐	☐ ☐	
	○	Significant and/or sensitive			
Number of project sites	○	2 or less	☐ ☐	☐ ☐	
	○	3 or more			
Impact on local environment	○	High	☐ ☐	☐ ☐	
	○	Low			

2 Project management					
Executive management involvement	○	Active involvement	☐ ☐	☐ ☐	
	○	Limited participation			
User management experience	○	Strong project experience	☐ ☐	☐ ☐	
	○	Weak project experience			
User management participation	○	Active participation	☐ ☐	☐ ☐	
	○	Limited participation			
Project manager	○	Experienced/full-time	☐ ☐	☐ ☐	
	○	Unqualified/part-time			
Project management techniques	○	Effective techniques used	☐ ☐	☐ ☐	
	○	Ineffective or not applied			
Client's experience of project type	○	Client has prior experience	☐ ☐	☐ ☐	
	○	First for client			

Part 1 appendices

LIVERPOOL JOHN MOORES UNIVERSITY
Aldham Roberts L.R.C.
TEL. 051 231 3701/3634

3 Project characteristics

Complexity	○ Reasonably straightforward	□ □ □ □
	○ Pioneering/new areas	
Technology	○ Proven and accepted methods and products	□ □ □ □
	○ Unproven or new	
Impact of failure	○ Minimal	□ □ □ □
	○ Significant	
Degree of organisational change	○ Minimal	□ □ □ □
	○ Significant	
Scope	○ Typical project phase or study	□ □ □ □
	○ Unusual phase or study	
Foundation	○ First phase or continuation	□ □ □ □
	○ Earlier work uncertain	
User acceptance	○ Project has strong support	□ □ □ □
	○ Controversy over project	
Proposed time	○ Reasonable allowance for delay	□ □ □ □
	○ Tight/rapid build-up	
Scheduled completion	○ Flexible with allowances	□ □ □ □
	○ Absolute deadline	
Potential changes	○ Stable industry/client/ application	□ □ □ □
	○ Dynamic industry/client/ application	
Work days (developer)	○ Less than 1,000	□ □ □ □
	○ 1,000 or more	
Cost-benefit analysis	○ Proven methods or not needed	□ □ □ □
	○ Inappropriate approximations/ methods	
Hardware/software capacity estimates	○ None or proven methods	□ □ □ □
	○ Unproven methods/ no contingency	

4 Project staffing

User participation	○ Active participation	□ □ □ □
	○ Limited participation	
Project supervision	○ Meets standards	□ □ □ □
	○ Below standards	
Project team	○ Adequate skills/experience	□ □ □ □
	○ Little relevant experience	

5 Project costs			
Cost quotation	○	Normal (i.e. time-based)	☐ ☐ ☐ ☐
	○	Fixed price	
Cost estimate basis	○	Detailed plan/proven method	☐ ☐ ☐ ☐
	○	Inadequate plan/method	
Formal contract	○	Non-standard form	☐ ☐ ☐ ☐
	○	Standard form	

6 Other

APPENDIX 10 Guidance to value management

Value management and value engineering

Value management (VM) and value engineering (VE) are techniques concerned with achieving 'value for money'.

VE was pioneered by an American, Lawrence Miles, during the Second World War to gain maximum function (or utility) from limited resources. It is a systematic team-based approach to securing maximum value for money where:

$$Value = Function/Cost$$

Thus value can be increased by improved function or reduced cost. The technique involves identification of high cost elements, determination of their function and critical examination of whether the function is needed and/or being achieved at lowest cost. In terms of projects, VE has the greatest influence and impact at the strategy/design stage. It requires reliable and appropriate cost data and uses brainstorming workshops by a group of experts under the direction of a facilitator.

VM is similar to VE but in terms of projects is focused on overall objectives and is most appropriate at the option identification and selection stage where the scope for maximising value is greatest.

Brainstorming forums involving those who would naturally contribute to the project and/or those with a significant interest in the outcome are a fundamental component of both VE and VM. Participants should be free to put up ideas and as far as possible idea generation and analysis should be kept separate. It is the role of the chairman to ensure that this is the case and hence good facilitation skills and a measure of independence are essential characteristics of the role.

The process

Value techniques are founded on three principal themes:

1. Achievement of tasks through involvement and teamwork, based on the premise that a team will almost always perform better than an individual.

2. Using subjective judgement, which may or may not incorporate risk assessment.

3. Value is a function of cost and utility in its broadest sense.

Key decisions in the application of value techniques are:

When should the technique be utilised?

Who should be involved?

Who should perform the role of the facilitator?

■ A balance must be struck between early application before an adequate understanding of the problem and constraints has been achieved and late application when conclusions have been drawn and opinions hardened. Although feasibility (when identifying suitable options) and pre-construction (before design freeze) would in most cases be suitable, each project should be examined on its merits.

■ The facilitator's role is to gain commitment and motivate participants, draw out all views and ensure a fair hearing, select champions to take forward ideas generated and keep to the agenda. To achieve these goals the facilitator must be independent, possess well-developed interpersonal and communication skills and be able to empathise with all participants. Although he/she must understand the nature of the project this need not be at a detailed level. Large, complex or otherwise difficult projects may warrant employment of an external specialist facilitator. The facilitator's role is crucial to the success of the exercise and care needs to be taken over selection.

■ The purpose and the agenda for the value forum should be determined by the project team. A value statement should also be produced giving a definition of value in relation to the particular project. For example, value may not be related solely to cost but may also encompass risk, environmental impact, occupational utility, etc. Although the significance of these factors will be project specific, the project team must ensure that this statement reflects corporate policy. The statement is not intended to be a constraint but is used as a benchmark throughout the forum to maintain focus.

■ Two to three days are usually required for each value forum. The project manager must ensure that all supporting information is available to the forum in summarised form and that expert advice is readily available. The project manager must therefore ensure that personnel with a detailed knowledge of the project participate in the forum.

Link to risk assessment

Value techniques may be used in conjunction with risk assessment where there is a variety of means of managing risk and choices have to be made. The process is particularly valuable in identifying the optimum mitigation approach where risk management options impinge on a variety of project objectives.

In this case risk management objectives are determined, in open forum, alongside overall project objectives. Risk management options are then ranked against the full range of objectives to determine the best option overall.

Potential pitfalls

1. Cost (monetary and time) of value meetings can be high.

2. At the feasibility/option identification stage aspects of the technique can conflict with the principles of the economic appraisal which seeks to identify an optimum solution by reference to an absolute measure of benefit as opposed to the subjective criteria used in value techniques, e.g. a standard of protection could not be specified as a value objective.

If benefits or costs could be assigned to all criteria, there would be no role for VM analysis at the feasibility stage although the facilitative and team-building aspects of the technique would still be useful.

3. Where economic appraisal overrides VM, team-building benefits will be undermined.

4. Client participation in the value process could prove to be detrimental in the event of dispute with a consultant/contractor. However, client participation is an integral part of the process, particularly at the feasibility stage, and must therefore be performed by experienced and knowledgeable staff aware of the contractual pitfalls.

Value techniques should be applied where there is a reasonable prospect of cost saving or substantial risk reduction or where consensus is necessary and difficult to achieve. Examples may include high value or complex projects impinging on a variety of interests and projects where environmental and/or intangible benefits are significant but difficult to quantify. Value techniques may be used whenever there is a need to define objectives and find solutions.

An example of utilisation of VM at key stages in a construction project framework is shown below.

Stage of study

	Inception/Feasibility	Strategy/Pre-construction	Construction

	VM 1	VM 2	VM 3	VM 4	VM 5
Milestone	Confirm concept	Select options and design approach	Detail design brief	Select constructors	Select components & commission
Workshop	Management issues	Technical issues	Design issues	Selection criteria	Details & processes
Objectives	Agree concept brief, reconcile views	Select options, design brief & procurement process	Maximum benefits, minimum costs, design brief operability	Buildability, documentation, best value contractors	Optimise components, validate processes

APPENDIX 11 Guidance to environmental impact assessment

Introduction

Environmental impact assessment (EIA) is a key instrument of European Union environmental policy. Since passage of the first EIA Directive in 1985 (Directive 85/337/EEC) both the law and the practice of EIA have evolved. An amending Directive was published in 1997 (Directive 97/11/EC).

Assessment of the effects of certain public and private projects on the environment is required under the Town and Country Planning Impact (Environmental Assessment) (England and Wales) Regulations 1999 (amended 2000) SI, in so far as it applies to development under the Town and Country Planning Act 1990.

EIA is a means of drawing together, in a systematic way, an assessment of a project's likely significant environmental effects. This helps to ensure that the importance of the predicated effects, and the scope for reducing them, are properly understood by the public and the relevant competent authority before it makes its decision.

Where EIA is required there are three broad stages to the procedure:

1 The developer must compile detailed information about the likely main environmental effects. To help the developer, public authorities must make available any relevant environmental information in their possession. The developer can also ask the 'competent authority' for their opinion on what information needs to be included. The information finally compiled by the developer is known as an 'environmental statement' (ES).

2 The ES (and the application to which it relates) must be publicised. Public authorities with relevant environmental responsibilities and the public must be given an opportunity to give their views about the project and ES.

3 The ES, together with any other information, comments and representations made on it, must be taken into account by the competent authority in deciding whether or not to give consent for the development. The public must be informed of the decision and the main reason for it.

The regulations

The regulations integrate the EIA procedures into the existing framework of local authority control. These procedures provide a more systematic method of assessing the environmental implications of developments that are likely to have significant effects. EIA is not discretionary. If significant effects on the environment are likely, EIA is required. Where the EIA procedure reveals that a project will have an adverse impact on the environment, it does not follow that planning permissions must be refused. It remains the task of the local planning authority to judge each planning application on its merits within the context of the Development Plan, taking account of all materials considerations, including the environmental impacts.

For developers, EIA can help to identify the likely effects of the a particular project at an early stage. This can produce improvements in the planning and design of the development; in decision making by both parties; and in consultation and responses thereto, particularly if combined with early consultations with the local planning authority and other interested bodies during the preparatory stages. In addition, developers may find EIA a useful tool for considering alternative approaches to a development. This can result in a final proposal that is more environmentally acceptable, and can form the basis for a more robust application for planning permission. The presentation of environmental information in a more systematic way may also simplify the local planning authority's task of appraising the application and drawing up appropriate planning conditions, enabling swifter decisions to be reached.

For EIA applications, the period after which an appeal against non-determination may be made is extended to 16 weeks.

Environmental impact assessment (EC Regulations)

EIA is a procedure required under the terms of European Union Directives on assessment of the effects of certain public and private projects on the environment.

Key stages	Notes
Project preparation	The client prepares the proposals for the project.
Notification to Competent Authority	The Competent Authority (CA) may be the Environment Agency, English Nature or similar organisations depending on the nature and the location of the project.
Screening	The CA makes a decision on whether EIA is required. This may happen when the CA receives notification of the intention to make a development consent application, or the developer may make an application for a screening opinion. The screening decision must be recorded and made public.
Scoping	The EC Directive provides that developers may request a scoping opinion from the CA. The scoping opinion will identify the matters to be covered in the environmental information. It may also cover other aspects of the EIA process.
Environmental studies	The developer carries out studies to collect and prepare the environmental information required.
Submission of environmental information to CA	The developer submits the environmental information to the CA together with the application for development consent. The environmental information is usually presented in the form of an environmental impact statement (EIS).
Review of adequacy of the environmental information	The client may be required to provide further information if the submitted information is deemed to be inadequate.

(cont'd overleaf)

Consultation with statutory environmental authorities, other interested parties and the public	The environmental information must be made available to authorities with environmental responsibilities and to other interested organisations and the general public for review. They must be given an opportunity to comment on the project and its environmental effects before a decision is made on development consent.
Consideration of the environmental information by the CA before making development consent decision	The environmental information and the results of consultations must be considered by the CA in reaching its decision on the application for development consent.
Announcement of decision	The decision must be made available to the public including the reasons for it and a description of the measures that will be required to mitigate adverse environmental effects.
Post-decision monitoring if project is granted consent	There may be a requirement to monitor the effects of the project once it is implemented.

Establishing whether EIA is required

Generally, it will fall to local planning authorities in the first instance to consider whether a proposed development requires EIA. For this purpose they will first need to consider whether the development is decribed in Schedule 1 or Schedule 2 to the Regulations. Development of a type listed in Schedule 1 always requires EIA. Development listed in Schedule 2 requires EIA if it is likely to have significant effects on the environment by virtue of factors such as its size, nature or location.

Development which comprises a change or extension requires EIA only if the change or extension is likely to have significant environmental effects.

Like the Town and Country Planning Act, the Regulations do not bind developments by Crown bodies.

Planning applications

Where EIA is required for a planning application made in outline, the requirements of the Regulations must be fully met at the outline stage since reserved matters cannot be subject to EIA. When any planning application is made in outline, the local planning authority will need to satisfy themselves that they have sufficient information available on the environmental effects of the proposal to enable them to determine whether or not planning permission should be granted in principle.

Where the authority's opinion is that EIA is required, but not submitted with the planning application, they must notify the applicant within 3 weeks of the date of receipt of the application, giving full reasons for their view clearly and precisely.

An applicant who still wishes to continue with the application must reply within 3 weeks of the date of such notification. The reply should indicate the applicant's intention either to provide an ES or to ask the Secretary of State for a screening direction. If the applicant does not reply within 3 weeks, the application will be deemed to have been refused.

Preparation and content of an environmental statement

It is the applicant's responsibility to prepare the ES. There is no statutory provision as to the form of an ES. However, it must contain the information specified in Part II, and such of the relevant information in Part I of Schedule 4 to the Regulations as is reasonably required to assess the effects of the project and which the developer can reasonably be required to compile. (See Appendix 3.)

The list of aspects of the environment which might be significantly affected by a project is set out in paragraph 3 of Part I of Schedule 4 (see Appendix 4), and includes human beings, flora, fauna, soil, water, air, climate, landscape, material, assets, including architectural and archaeological heritage, and the interaction between any of the foregoing.

Procedures for establishing whether or not EIA is required ('screening')

The determination of whether EIA is required for a particular development proposal can take place at a number of different stages.

i. The developer may decide that EIA will be required and submit a statement.

ii. The developer may, before submitting any planning application, request a screening opinion from the local planning authority. If the developer disputes the need for EIA (or a screening opinion is not adopted within the required period), the developers may apply to the Secretary of State for a screening direction. Similar procedures apply to permitted development.

iii. The local planning authority may determine that EIA is required following receipt of a planning application. If the developer disputes the need for EIA, the applicant may apply to the Secretary of State for a screening direction.

iv. The Secretary of State may determine that EIA is required for an application that has been called in for his or her determination or is before him or her on appeal.

v. The Secretary of State may direct that EIA is required at any stage prior to the granting of consent for a particular development.

Provision to seek a formal opinion from the local planning authority on the scope of an ES ('scoping')

Before making a planning application, a developer may ask the local planning authority for their formal opinion on the information to be supplied in the ES (a 'scoping opinion'). This provision allows the developer to be clear about what the local planning authority considers the main effects of the development are likely to be and, therefore, the topics on which the ES should focus.

The developer must include the same information as would be required to accompany a request for a screening opinion.

The local planning authority must adopt a scoping opinion within 5 weeks of receiving a request.

Provision of information by the consultation bodies

Under the Environmental Information Regulations, public bodies must make environmental information available to any person who requests it. Once a developer has given the local planning authority notice in writing that they intend to submit an ES, the authority must inform the consultation bodies.

The consultation bodies are:

i. The bodies who would be statutory consultees under Article 10 of the GDPO for any planning application for the proposed development.

ii. Any principal council for the area in which the land is situated (other than the local planning authority).

iii. English Nature.

iv. The Countryside Commission.

v. The Environment Agency.

Selection criteria for screening Schedule 2 development

This is a reproduction of Schedule 3 of the Regulations (paragraphs 20 and 33).

1. **Characteristics of development**

   The characteristics of development must be considered having regard, in particular, to:

   i. the size of the development

   ii. the cumulation with other development

   iii. the use of natural resources

   iv. the production of waste

   v. pollution and nuisances

   vi. the risk of accidents, having regard in particular to substances or technologies used.

2. **Location of development**

   The environmental sensitivity of geographical areas likely to be affected by development must be considered, having regard, in particular, to:

   i. the existing land use

   ii. the relative abundance, quality and regenerative capacity of natural resources in the area

   iii. the absorption capacity of the natural environment, paying particular attention to the following areas:

      a. wetlands

      b. coastal zones

      c. mountain and forest areas

    d.   nature reserves and parks

    e.   areas classified or protected under Member States' legislation; areas designated by Member States pursuant to Council Directive 79/409/EEC on the conservation of natural habitats and of wild fauna and flora

    f.   areas in which the environmental quality standards laid down in Community legislation have already been exceeded.

    g.   densely populated areas

    h.   landscape of historical, cultural or archaeological significance.

3.   **Characteristics of the potential impact**

The potential significant effects of development must be considered in relation to criteria set out under paragraphs 1 and 2 above, and having regard in particular to:

    i.   the extent of the impact (geographical area and size of the affected population)

    ii.   the transfrontier nature of the impact

    iii.   the magnitude and complexity of the impact

    iv.   the probability of the impact

    v.   the duration, frequency and reversibility of the impact.

Information to be included in an environmental statement

This is a reproduction of Schedule 4 of the Regulations (paragraphs 81–85 and 91).

Part I

1.   Description of the development, including in particular:

    i.   a description of the physical characteristics of the whole development and the land-use requirements during the construction and operational phases

    ii.   a description of the main characteristics of the production processes, for instance, nature and quantity of the materials used

    iii.   an estimate, by type and quantity, of expected residues and emissions (water, air and soil pollution, noise, vibration, light, heat, radiation etc) resulting from the operation of the proposed development.

2.   An outline of the main alternatives studied by the applicant or appellant and an indication of the main reasons for his choice, taking into account the environmental effects.

3.   A description of the aspects of the environment likely to be significantly affected by the development, including, in particular, population, fauna, flora, soil, water, air, climatic factors, material assets, including the architectural and archaeological heritage, landscape and the inter-relationship between the above factors.

4.   A description of the likely significant effects of the development on the environment, which should cover the direct effects and any indirect, secondary,

cumulative, short, medium and long term, permanent and temporary, positive and negative effects of the development, resulting from:

   i.   the existence of the development

   ii.   the use of natural resources

   iii.   the emission of pollutants, the creation of nuisances and the elimination of waste

and the description by the applicant of the forecasting methods used to assess the effects on the environment.

5.   A description of the measures envisaged to prevent, reduce and where possible offset any significant adverse effects on the environment.

6.   A non-technical summary of the information provided under paragraphs 1 to 5 of this Part.

7.   An indication of any difficulties (technical difficulties or lack of know-how) encountered by the applicant in compiling the required information.

Part II

1.   A description of the development comprising information on the site, design and size of the development.

2.   A description of the measures envisaged in order to avoid, reduce and, if possible, remedy significant adverse effects.

3.   The data required to identify and assess the main effects which the development is likely to have on the environment.

4.   An outline of the main alternatives studied by the applicant or appellant and an indication of the main reasons for his choice, taking into account the environmental effects.

5.   A non-technical summary of the information provided under paragraphs 1 to 4 of this part.

The characteristics of a good EIS

- Has a clear structure with a logical sequence, for example, describing existing baseline conditions, predicted impacts (nature, extent and magnitude), scope for mitigation, agreed mitigation measures, significance of unavoidable/residual impacts for each environmental topic.

- Has a table of contents at the beginning of the document.

- Has a clear description of the development consent procedure and how EIA fits within it.

- Reads as a single document with appropriate cross-referencing.

- Is concise, comprehensive and objective.

- Is written in an impartial manner without bias.

- Includes a full description of the development proposals.

- Makes effective use of diagrams, illustrations, photographs and other graphics to support the text.

■ Uses consistent terminology with a glossary.

■ References all information sources used.

■ Has a clear explanation of complex issues.

■ Contains a good description of the methods used for the studies of each environmental topic.

■ Covers each environmental topic in a way which is proportionate to its importance.

■ Provides evidence of good consultations.

■ Includes a clear discussion of alternatives.

■ Makes a commitment to mitigation (with a programme) and to monitoring.

■ Has a non-technical summary which does not contain technical jargon.

Example layout of an environmental statement

1.0 INTRODUCTION

1.1 EIA Development
1.2 Planning Context
1.3 About ???
1.4 Scope and Content of the ES
1.5 ES Availability and Comments

2.0 EIA METHODOLOGY

2.1 Objectives
2.2 Scoping Study
2.3 Consultations
2.4 Defining the Baseline
2.5 Sensitive Receptions
2.6 Impact Prediction
2.7 Evaluation of Significance
2.8 Mitigation
2.9 Residual Impact
2.10 Assumptions and Limitations

3.0 DEVELOPMENT BACKGROUND AND ALTERNATIVES

3.1 Introduction
3.2 Site Considerations and Constraints
3.3 No Development Alternative
3.4 Objectives of the Proposed Redevelopment
3.5 Design Alternatives

4.0 THE SITE DESCRIPTION AND DESIGN STATEMENT

4.1 Introduction
4.2 Site Location and Setting
4.3 Site Description
4.4 The Design Statement

5.0 PLANT DISMANTLING, DEMOLITION, REMEDIATION AND CONSTRUCTION

5.1 Introduction
5.2 Schedule Overview
5.3 Plant Dismantling and Asbestos Removal
5.4 Demolition
5.5 Remediation
5.6 Construction
5.7 Construction Traffic
5.8 Environmental Management Plan and Code of Construction

6.0 ENVIRONMENTAL MANAGEMENT PLAN AND POTENTIAL IMPACTS OF THE CONSTRUCTION WORKS

6.1 Introduction
6.2 Scope of the EMP
6.3 Summary and Conclusions

7.0 PLANNING AND POLICY CONTEXT

7.1 Introduction
7.2 Planning Policy Guidance
7.3 Strategic Guidance
7.4 Strategic Planning in (location)
7.5 Planning Brief
7.6 The Adopted RBKC UDP (August 1995)
7.7 RBKC UDP Proposed Alterations
7.8 LBHF Adopted UDP
7.9 LBHF Revised Deposit Draft (June 2000)
7.10 Affordable Housing
7.11 Summary and Conclusions to Planning and Policy Context

8.0 SUSTAINABILITY

8.1 Introduction
8.2 National Guidance and Local Policy
8.3 Approach to Assessment
8.4 Sustainability Topics
8.5 Results
8.6 Summary and Conclusions

9.0 SOCIO-ECONOMIC IMPACTS

9.1 Introduction
9.2 Approach to Assessment
9.3 Baseline Statistics
9.4 Impact of the Development
9.5 Summary and Conclusions

10.0 BUILT HERITAGE, TOWNSCAPE AND VISUAL IMPACTS

10.1 Introduction
10.2 Approach to Assessment
10.3 Approach to Presentation of the Visual Assessment
10.4 Baseline Condition – The Heritage and Existing Townscape
10.5 Townscape Studies, Policies and Guidelines
10.6 Townscape and Visual Impact Assessment
10.7 Impact Assessment – The Immediate Locality
10.8 Impact Assessment – The Panorama, The Moving Eye and the Interaction of Major Built Forms
10.9 Impact Assessment – Specific Viewpoints Reviewed
10.10 Summary and Conclusions

11.0 ARCHAEOLOGY

11.1 Introduction
11.2 Approach to Assessment
11.3 Policy Considerations and Legislative
11.4 Initial Assessment
11.5 Archaeological Potential of the Site
11.6 Environmental Potential of the Site
11.7 Archaeological Resources in the Surrounding Area
11.8 Summary of Archaeological Potential
11.9 Impact of the Development

APPENDIX 12 Application of project management software

There is a wide range of project management software packages available, however, there is not a single package which is ideal for projects and budgets. If the requirements are simple and mainly scheduling is needed, and the project does not require an analysis of the associated resources, then there is an unnecessary overhead, both in terms of cost and also in the complexity of the user interface, associated with the more complicated packages. In such projects perhaps a simpler package will be ideal. Such a package can be used collaboratively over the Internet (with both Macs and PCs) and can be customised if necessary.

For professionals who use project planners daily, keyboard shortcuts are often significantly more efficient than using a mouse. For such users, exporting and importing data from databases are also key priorities and sophisticated report generation is essential. There is a wide range of products in this area.

Perhaps the most widely used package is Microsoft Project. This software places a significant emphasis on updates via e-mail and close integration with the web and Office applications. Because it is so widely used there is also a developing market for add-on software for this package.

The market for project management software has traditionally been directed at large businesses with large budgets and packages are often very expensive. But a number of companies are now addressing the mass market with relatively powerful, easy-to-use products, at a price more usually associated with home or small office software.

Basic features of project management software

As with all software the types and extent of features and capabilities vary with different packages. Some of the basic features usually found in project management software are:

- Front-end modelling, for planning and estimating the size, risk and overall architecture of the project.

- Time reporting capability, for analysing trends against base values and current plans.

- Work request system, for tracking and completing requested tasks within an acceptable time.

- Project accounting, for linking time reporting to specific projects to be charged for work, and for linking the project changes to the organisation's accounting system.

- Resource levelling, for efficiently reviewing work request records and comparing those data with other factors that change resource availabilities constantly.

- Integration, for co-ordinating the functions of various aspects of software, which is important because of the complex mix of projects and other activities occurring.

Utilities of project management software

Project management software can be used for planning, scheduling, monitoring and controlling all aspects of the project. The concept of project management involves dividing the overall project into individual activities or work units. For each activity, the following information is then recorded:

- requirements for successful completion of the activity (project constraints)

- the earliest time the activity could begin (earliest start)

- the earliest time it could end (earliest finish)

- any 'slack time', or time that the activity could be delayed without delaying the entire project (float).

Information of this type is then used as a basis for managing the entire project.

Applications of project management software

Some of the uses of project management software include:

- Determining resource requirements and assigning resource to tasks.

- Locating potential problems (e.g. conflicts) in the work schedule and recommending solutions.

- Maintaining accounting records of the project.

- Preparing management reports concerning the project.

Project management software can be used effectively for planning before the actual project begins. During the implementation of the project, the software is useful for displaying progress reports or for indicating changes that need to be made. Most projects include some unexpected occurrences (variations/changes), and one the of the biggest advantages of project management software is the ease of making changes to the data to determine what effect any changes would have on the overall results or time schedule. The process of correctly deciding on an appropriate option is perhaps much better informed as several options can be tested and a good picture of the overall results of each choice can be obtained to facilitate decision making.

APPENDIX 13 Change management

Change in a construction project is any incident, event, decision or anything else that affects:

- The scope, objectives, requirements or brief of the project.

- The value (including project cost and whole-life cost) of the project.

- The time milestones (including design, construction, occupation).

- Risk allocation and mitigation.

- Working of the project team (internally or externally).

- Any project process at any project phase.

Changes during the design development process

The procedure outlined is used to control the development of the project design from the design brief to preparation of tender documents. It will include:

- addressing issues in the design brief

- variations from the design brief, including design team variations and client variations

- developing details consistent with the design brief

- approving key design development stages, namely scheme design approval and detailed design approval.

The procedure is based on the design development control sheet. The approved design will comprise the design brief and the full set of approved design development control sheets.

The procedure comprises the following stages:

- The appropriate member of the design team addresses each design issue in the development of the brief, co-ordinated by the design team leader.

- Proposals developed are discussed with the appropriate members of the project's core group through submission of detailed reports/meetings co-ordinated by the project manager. Reports should not repeat the design brief, but expand it, address an issue and prepare a change.

- The design team leader co-ordinates preparation of a design development control sheet, giving:

  ○ design brief section and page references

  ○ a statement of the issue

  ○ a statement of the options

  ○ the cost plan item, reference and current cost

  ○ the effect of the recommendation on the cost plan and the schedule

  ○ a statement as to whether the recommendation requires transfer of client contingency (i.e. a client variation to the brief) and if so the amount to be transferred.

- The design team report section of the control sheet is signed by:
    - o   the design team member responsible for recommendations
    - o   the quantity surveyor (for cost effect)
    - o   the design team leader (for co-ordination).

- The design team leader sends the design development control sheet to the project manager who obtains the client's approval signature and returns it to the design team leader.

- The quantity surveyor incorporates the effect of the approved recommendation into the cost plan.

- The project manager incorporates the effect of the approved recommendation into the master schedule.

Design development control sheet

Client name:	
Project name:	
Sheet no.	

Design team report

Design brief section:

Issue: Pages:

Options considered: 1.

                           2.

                           3.

Recommendation:

Cost plan item:

Ref:

Current cost:

Effect of recommendation on costs/schedules: Increase/decrease

Application for transfer of client contingency: Yes/No   Amount:

Architect/services engineer/structural engineer: Date:

Quantity surveyor: Date:

Design Team leader: Date:

Client approval

Design development/Client contingency transfer approved (delete as applicable)

Position   Signature   Date

Part 1 appendices

Example of change management process

1. Identification of requirement for change.

2. Evaluation of change.

3. Consideration of implications and impact including risks.

4. Preparation of change order.

5. Review of change order – client's decision stage.

6. Implementation of change.

7. Feedback including causes of change.

Change order request form

Project no.	Date:	No.	
Client: **Project:**			**Distribution:**
Subject – definition of change:			WHAT
Identified by:			WHO
Reasons for change:			WHY
	Discretionary		Non-discretionary	
Cost implication:			
Time implication:			
Recommended action:	Project manager		Date
Client decision required by:	Date:		
Forwarded to client:	Date:		
Client's decision:	Date:		
Projected schedule and cost plan (budget) amended on	Project manager		

Change order register

Request Id:	Date	Initiated by:	Description of Change		Client Decision required by:	Client Decision obtained by:	Client Decision	Client Decision Id no.:
Project:			**Client:**	**Job Id:**		**File Reference:**		

Sources and further information: (1) Project Management Research Team, CII (1994) *Project Change Management*, Construction Industry Institute, Special Publication 43-1; (2) CIRIA (2001) *Managing Project Change – A Best Practice Guide*, Lazarus, D and Clifton, R, CIRIA C556.

APPENDIX 14 Procedure for the selection and appointment of consultants

Stages	Key steps
Strategy	■ Decide works procurement strategy
	■ Prepare project brief
	■ Prepare consultant's brief
	■ Decide terms of engagement including the choice of single/multiple appointment and phased appointment
Pre-selection	■ Prepare preliminary list
	■ Decide criteria of selection
Selection	■ Invite to tender
	■ Evaluate tender
	■ Assess tender
Appointment	■ Finalise terms of engagement
	■ Finalise management, monitoring and review process

Guidance for selection process

1. Determine what duties are to be assumed by the consultants and prepare a schedule of responsibilities. If applicable, consider what level of in-house expertise is available.

2. Check to see if the client has any in-house procedures or standard conditions of engagement for the appointment of consultants and what scope there is for deviating from them.

3. Decide on the qualities most needed for the project, and the method of appointment. Agree them with the client.

4. Establish criteria for evaluating consultants with weighting values (e.g. 5 vital, 0 unimportant) for each criterion (see Appendix 12).

5. Assemble a list of candidates from references and recommendations. Check any in-house approved and updated lists of consultants.

6. Prepare a short list by gathering information about possible candidates. Check which firms or individuals are prepared in principle to submit a proposal.

7. Assess candidates against general criteria and invite proposals from a select number (no more than six and no fewer than three per discipline). Invitation documents should be prepared in accordance with the checklist given below. Competitive fee bids, if required, should conform to relevant codes of practice.

8. Arrange for conditions of engagement to be drawn up. The conditions, the form of which will vary with the work required and the type of client, should refer to a schedule of responsibilities for the stages for which the consultant is appointed and include a clause dictating compliance with the project hand-book. The conditions of engagement should be based as closely as possible

on industry standards (e.g. as set by RIBA, ACE, NEC and RICS). Consistency of style and structure between conditions for different members of the team will improve each member's understanding of their own and others' responsibilities. Each set should include this aspect. Fee calculation and payment terms should be clearly defined at the outset, together with the treatment of expenses, i.e. included or not in the agreed fee.

9. Determine the criteria for assessing the consultants' proposals. Agree them with the client.

10. Appraise the proposals and select the candidates most appropriate for the project. Proposals should be analysed against the agreed criteria using weighting analysis.

11. Arrange final interviews with selected candidates (minimum of two) for final selection/negotiation as necessary.

12. Submit a report and recommendation to the client.

13. Client appoints selected consultant.

14. Unsuccessful candidates are notified that an appointment has been made.

Checklist

1. The consultant's brief must include:
   - project objectives
   - requirements of other participants
   - services to be provided
   - project schedule including the key dates
   - requirement of reports including key dates.

2. Invitation documents must include:
   - a schedule of responsibilities
   - the form of interview panel
   - draft conditions of engagement (an indication of the type to be used)
   - design skills or expertise required
   - personnel who will work on the project, their roles, time-scales, commitment and output
   - warranties required, for whose benefit and in what terms.

Invitations should ask candidates to include information on the level of current professional indemnity insurance cover for the duration of the project. Details of policy, date of expiry and extent of cover for subcontracted services must be provided.

Example of consultancy services at different project stages

The following is a list of consultancy services typically provided at different stages of a project. However, this neither is a comprehensive list nor outlines the preferred sequence as that may vary from project to project.

Inception/feasibility

- Identification of client requirements and objectives including preparation of the project brief.

- Feasibility studies including evaluation of options, environmental impact assessment, site assessment, planning guidance and commercial assessment.

Strategy/pre-construction

- Design development including preparation of outline design and scheme design.

- Development of cost estimates, tender preparation and evaluation and preparing project schedule.

- Preparation of construction specifications and schedules.

Construction/commissioning

- Preparation and issuing working drawings and variations.

- Project/construction management.

- Inspection, monitoring and valuation of construction.

- Certification of payment.

- Advising on dispute resolution.

- Confirmation of completion.

- Assisting in project handover.

Completion/handover

- Ensuring defects correction.

- Settlement of project including final accounts.

- Confirmation of operation and maintenance procedures.

- Post-project appraisal and feedback.

Source and further information: Association of Consulting Engineers (1993) *Balancing Quality and Price*; CIRIA (1994) *Value by Competition*; OGC (2002) *Guide to the Appointment of Consultants and Contractors*.

APPENDIX 15 Characteristics of different procurement options

The significance of the features listed in Table 3.1 may be outlined as follows.

1. Having widely dispersed responsibilities for different activities may provide the project manager with greater control, e.g. in the selection of preferred consultants. It may, however, make it difficult to pinpoint responsibility.

2. It is acceptable practice to limit the numbers of tenders invited from contractors, based on value and design development criteria. Where tendering involves significant design development, the cost will discourage contractors, unless invitations are restricted. Such restrictions may not produce the most competitive price available, unless careful pre-tender assessments are made.

3. Although the establishment of a certain financial outcome at an early stage in the development schedule will minimise client risks, it could well be at a price. This is because of the risks which the tenderers will have to assume. A balance has to be achieved that depends on all the circumstances.

4. The client's requirements document associated with design and build procurement is a definitive statement. It must be produced early and it becomes the basis for all subsequent activities. Other procurement options enable progressive development of the client's brief, which may be helpful where there is uncertainty or greater complexity.

5. Independent assistance with the development of a design brief, which is integral to procurement options, may be advantageous where there is uncertainty or greater complexity, similar to item 4.

6. Mobilisation of construction using traditional procurement is relatively slow because much of the design development must be completed before appointment of the contractor, whereas all other methods enable progressive design and construction.

7. Little flexibility to accommodate variations exists within the design and build method. The other methods make reasonable provision for flexibility through the issuing of variations or additional works contracts.

8. Standard documentation with which the industry is familiar allows agreements to be entered readily. Although it enables incorporation of particular requirements the drafting of unique documents often involves much negotiation and expense.

9. Where there are significant uncertainties or where limited finance is available, the opportunity to develop and appraise proposals may be advantageous. There may even be an opportunity to carry out construction on a progressive basis, step by step.

10. All procurement methods should seek to provide facilities for client cost monitoring, although the detail may vary.

11. Contractors' input to design could produce more cost-effective solutions provided the contractors' interests are accommodated correctly. By using design and build, the contractor clearly has a vested interest in providing such input.

12. The schedule for preparation of production information is often critical to, and should be determined by, the construction schedule.

13. Procurement methods have different abilities to select preferred trade or works contractors that actually execute the works; no influence in selection is

possible using design and build, and only limited influence is possible using the traditional method.

14. Design and build procurement makes no provision for the monitoring of construction quality – any monitoring required by the client must be independently commissioned. In other forms of procurement members of the design team – or the management contractor or the construction manager – may have monitoring responsibilities. But in all cases, except the last, only limited control of quality is available.

15. Since construction works involve substantial financial transactions, there is considerable financial benefit to the main contractor in achieving payments as promptly as possible while delaying payments due as long as possible. This may have a significant detrimental effect on the attitudes and performance of the specialist subcontractors, and hence on the quality of their workmanship, thus exacerbating the limited quality control characteristics of procurement methods. Where payments to specialist contractors are under direct control of the client/construction manager, this can be turned to advantage.

16. Management procurement methods provide for remuneration of the management contractor or construction manager on the basis of a fee, not necessarily related to performance. Equitable performance measurement is often difficult. In design and build procurement there is a strong incentive for good management; in traditional procurement there is also an incentive for the contractor.

17. Construction quality, speed and cost can all be improved through good teamwork. Procurement methods which recognise the varying responsibilities of those managing construction operations and which preclude exploitation of any party are most likely to avoid confrontation.

Selection of procurement method

From the foregoing it can be seen that the most important characteristics of each method will best suit particular types of project. For example, design and build procurement would be an obvious choice where a client has limited interest in involvement with the design or construction process; and when there are clearly defined, straightforward requirements, including a need for early determination of cost.

It is necessary to consider all the characteristics of the project and to compare them with the characteristics of the various procurement methods available. The most important characteristics should be identified initially, after which secondary and peripheral issues should be considered, and the details determined for any necessary adaptation of the basic procurement methods available. For example, although design and build procurement may appear well suited to the project characteristics, it will probably be appropriate for the client to appoint an architect or planning consultant to progress the project through planning approval. The documentation produced would then be incorporated in the client's requirements upon which design and build tenders would be sought.

Care must be taken in adapting any particular procurement method to compensate for perceived shortcomings, to avoid compromising the basic principles and essential characteristics. Thus, for example, although the engagement of design assistance for preparation of the client's requirements will inevitably dilute the single-point responsibility attribute of design and build procurement, the effects of this dilution should be mitigated by careful definition of responsibilities and terms of engagement. Similar care must be exercised when procuring specialist components or services incorporating design elements, within a traditional project procurement method.

Selection of a procurement method is thus an essential element in the development of the policies to be adopted for implementation of all projects. In view of the fundamental differences in philosophy between the four basic procurement methods, the procurement method should be determined at the earliest possible stage so that timely decisions can be made on the engagement of appropriate project resources. The development process can be optimised only by giving consideration at the earliest stage to the issues upon which the appropriate procurement method should be determined.

Identification of priorities

			Traditional	Design & Build	Management	Construction Management
Timing	Importance of early completion	Crucial	✓	✓	✓	✓
		Important	✓	✓	✓	✓
		Low	✓	✗	✗	✗
Controllable changes	Probability of variations	High	✓	✗	✓	✓
		Low	✗	✓	✗	✗
Technical complexity	Importance of advanced technology	High	✓	✗	✓	✓
		Medium	✓	✓	✓	✓
		Low	✗	✓	✗	✗
Price certainty	Importance of end-price fixation	High	✓	✓	✓	
		Low	✗	✓	✗	✓
Competition	Importance of competitive procurement	For all construction work	✓	✓	✓	✓
		Construction/management	✓	✓	✓	
		Not so important	✓	✓	✗	✗
Management	Capability of managing multiple consultants and contractors – as against appointing a single firm responsible for the project	Can manage separate firms	✓	✗	✓	✓
		Must have only one firm for everything	✗	✓	✗	✗
Technical responsibility	Importance of direct professional responsibility from the designers and cost consultants	High	✓	✗	✓	✓
		Low	✗	✓	✗	✗
Risk attitude	Nature of risk strategy	Retaining risks	✗	✗	✗	✓
		Sharing risks	✓	✗	✓	✗
		Delegating risks	✗	✓	✗	✗

Source: *Thinking about Building*, NEDO/HMSO (1985).

Selecting a procurement route

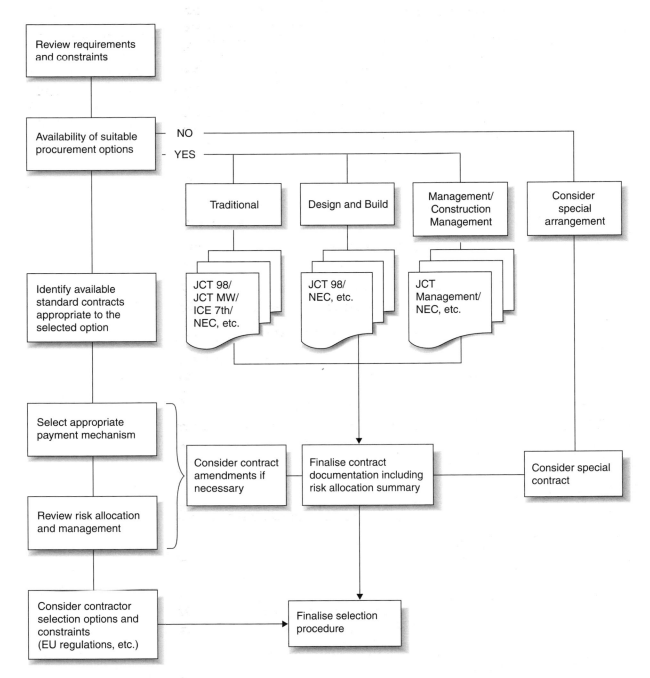

Source: CIRIA C556: *Managing Project Change*.

APPENDIX 16 Dispute resolution methods

The options for resolving disputes arising out of the contract should be considered at the strategy stage while selecting the procurement and contract strategy.

The parties in dispute necessarily would prefer a dispute resolution method having perhaps some or all of the following characteristics:

1. Fairness and natural justice in decisions.

2. Speed of process and arriving at a decision.

3. Cost effectiveness.

4. Establishment of a principle.

5. Common sense decisions.

Litigation

Litigation is available unless it is expressly excluded in the contract or by virtue of an arbitration clause. Depending on the nature of the claim and/or the sums involved, an application is made to the County Court or the High Court. In certain circumstances, construction contract claims can be referred to the Technology and Construction Courts (TCC).

Court proceedings are expensive and lengthy. Decisions of the lower courts are subject to appeal on both facts and law, which may delay settlement for a considerable time.

Arbitration

Arbitration is a legally binding process which can be invoked by consensual agreement between the parties. Party autonomy prevails and there is little judicial intervention by the courts.

Arbitration can be expressed within the terms of the contract or the parties can agree to arbitration should the contract be silent on the matter. In the UK arbitration is governed by the Arbitration Act 1996 and often specific Rules (such as CIMAR). Internationally, there is the International Chamber of Commerce (ICC) in Paris and the London Court of International Arbitration (LCIA). These institutions have published their own rules of procedure.

An arbitrator can be appointed by an appointing body or can be agreed by the parties and is chosen for his or her expertise in the subject matter of the dispute. Arbitrators are not specific to one discipline.

Arbitration is conducted in private and is confidential. No one other than those involved are entitled to attend a hearing. The arbitrator's award may itself be confidential unless it is subject to appeal in the courts. The arbitrator's jurisdiction (that is, what the arbitrator can decide) is governed by the arbitration agreement, generally the dispute resolution clause contained in the contract. The major technical institutions, such as the Chartered Institute of Building, the Law Society and the Chartered Institute of Arbitrators, all provide an appointment service.

The right of appeal is limited. There is no right of appeal on a finding of fact and a limited right concerning points of law.

Adjudication

The recent changes in legislation (Housing Grants, Construction and Regeneration Act 1996) have introduced adjudication as the normal form of dispute resolution for construction contracts, including the design of, carrying out and giving advice on construction operations. The Act gives the parties the right to an adjudication procedure compliant with the Act. The Act sets out eight principles, which if incorporated into the contract provide a procedure that satisfies its requirements. If not incorporated into the contract then the right is exercisable in accordance with a procedure set out in the Scheme for Construction Contracts Regulations 1998. Statutory adjudication is a 28-day procedure, extendable by agreement. An adjudicator can be appointed by agreement or by an Adjudicator Nominating Body (ANB). Examples of ANBs are: the Chartered Institute of Building, the Institution of Civil Engineers and the Chartered Institute of Arbitrators.

Several bodies publish adjudication procedures complying with the Act and tailored to particular branches of the industry. The adjudicator's decision is not final and can be summarily enforced in the courts. Thus, the decision is temporarily binding upon the parties until the dispute is finally settled by arbitration or litigation.

Alternative dispute resolution methods

A number of dispute resolution methods, as alternatives to the traditional routes of litigation and/or arbitration can be considered in order to save time and cost by providing a commercial solution which is aimed to be mutually acceptable. Some of these methods are listed below.

Mediation

Mediation is a process conducted by a 'mediator', usually trained by CEDR (Centre for Dispute Resolution) or similar organisations. The mediator does not have to be a lawyer. The process is started through a brief written summary of the case presented to the mediator in advance. Oral representations are used to clarify issues and search for the best possible areas of agreement or a settlement. The whole process is confidential, undertaken without prejudice but legally non-binding, and thus dependent upon the intent of the parties.

Conciliation

This process has been incorporated by the ICE forms of contract (6th edn onwards) as a route to dispute resolution. The conciliator usually puts forward a recommendation in order to resolve the dispute.

Expert evaluation

This constitutes appointing an expert through a contract to determine certain issues. The decision of the expert is binding upon the parties. However, if the contract does not provide for the appointment of an expert, in case of a dispute the parties may still decide to appoint an expert and it is up to the parties to agree whether the expert decision is going to be binding or not. Expert evaluation is perhaps most appropriate where the dispute concerns a particular expertise, such as valuation of variations, determining reasonable value, interpretation of technical specification, or assessing the reasonableness of an action taken by a technical expert (particularly on negligence or duty of care issues).

APPENDIX 17 Regular reports to the client

Notes for guidance on contents

Executive summary

The purpose of the executive summary is to give the client a snapshot of the project on a particular date which can be absorbed in a few minutes. It should contain short precise statements on the following:

(a) Significant events that have been achieved.

(b) Significant events that have not been achieved and action being taken.

(c) Significant events in the near future, particularly where they require specific action.

(d) Progress against the master, design and construction schedules.

(e) Financial status of the project.

Contractual arrangements (including legal agreements)

Each project requires the client to enter into a number of legal agreements with parties such as local authorities, funding institutions, purchasers, tenants, consultants and contractors.

The report should be subdivided to identify each particular agreement and to provide details of requirements and progress made against the original project master schedule.

The following are indications of possible legal agreements that may be required on a project:

- joint development agreement

- land purchase agreement

- funding agreement

- purchase agreement

- tenant/lease agreement

- consultants' appointments

- Town and Country Planning Acts: sections in force at the time, e.g.

  ○ planning gain

  ○ highways agreement

  ○ planning notices

  ○ land adoption agreement

  ○ public utilities diversion contracts.

Client's brief and requirements

This provides a 'status' report on how the client's brief and requirements are progressing. The report should identify any requirements which need clarification or amplification and also those which are still to be defined by the client.

Client change requests

Client-orientated changes should be listed under status (being considered, in progress, completed), cost and schedule implications. The objective is to make the client fully aware of the impact and progress of any change.

Planning, Building Regulations and fire officer consents	This section will be subdivided into the various consents required on a specific project. Each section should highlight progress made, problems, possible solutions and action required or in progress. The following are examples of possible consents:

- planning – outline
- planning – detailed, including conditions
- Building Regulations
- means of escape
- English Heritage/Historic Buildings
- fire officer
- public health
- environmental health
- party-wall awards.

Public utilities	Each separate utility should be dealt with in terms of commitment, progress, completion and any agreements and way-leaves as appropriate.
Design reports – summaries	The design team and consultants should prepare reports on progress, problems and solutions which will form the appendices and must include marked-up design schedules and 'issue of information' schedules.
	The design report, however, should be distilled into an 'impact-making' synopsis and agreed as a fair representation by each member concerned.
Health and safety	Report on the preparation of the CDM health and safety plan and the health and safety file.
Project master schedule	Updated schedules should form an appendix to the report, specifying progress made. A short commentary on any noteworthy aspects should be made under this section.
Tendering report	This is a status report on events leading up to the acceptance of tenders. It should show clearly how the various stages are progressing against the action plan.
Construction report summary	This report is prepared in a similar way to that outlined for design reports (above).
Construction schedule	The updated schedule should form an appendix to the report, highlighting progress made and showing where delays are occurring or are anticipated.
	A short commentary on any important items should be given in the report under this section.
Financial report	A fully detailed financial report should form one of the appendices. It should provide a condensed overview (say two to three pages) giving the financial status and cash flow of the project. This report will embrace the information provided by the quantity surveyor and also call for the project manager to provide an overall financial view, highlighting any specific matters of interest to the client.
Appendices	These will include full reports and schedule updates as outlined in the previous sections. Other reports, possibly of a specialist nature, may also be included.
	Should the report be presented at a formal meeting then the minutes of previous meetings should be included in the appendices.

APPENDIX 18 Practical completion checklist

Project no:

Project Management Ltd

Authorised to approve   Signature

_____   _____

_____   _____

Have/has the following been completed?

1. Contract works. ☐
2. Commissioning of engineering services. ☐
3. Outstanding works schedule issued. ☐
4. Outstanding works completed. ☐
5. Operating and maintenance manuals, 'as built' drawings and C&T records issued ☐
6. Maintenance contracts put into place. ☐
7. Building Regulations consent signed off. ☐
8. Occupation certificate issued. ☐
9. Public health consent signed off. ☐
10. Health and safety consent signed off and health and safety file available. ☐
11. Planning consent complied with in full, including reserved matters. ☐
12. Equipment test certificates issued (lifts, cleaning cradle, others). ☐
13. Insurers' certificates issued (lifts, cleaning cradle, sprinklers, others). ☐
14. Means of escape signed off. ☐
15. Fire-fighting systems and appliances signed off. ☐
16. Fire alarm system signed off and fire certificate issued. ☐
17. Public utilities way-leaves and lease agreements signed off. ☐
18. Public utilities supplies inspected and signed off. ☐
19. Licences to store controlled chemicals. ☐
20. Licences to dispose of controlled chemicals. ☐
21. Licences to store gases. ☐
22. Licence to use artesian well. ☐
23. Adoption of highways, estate roads and walkways by local authorities. ☐
24. Consent to erect and maintain flag-poles. ☐
25. Consent to erect illuminated signs. ☐
26. Cleaning to required standard. ☐
27. Removal of unwanted materials and debris. ☐
28. Tools and spares. ☐
29. Client/user insurances established. ☐

Completed   /

Not applicable   ×

APPENDIX 19 Facilities management

Facilities management started out as property management, concerned primarily with the management of premises. As commercial reality and competitiveness demanded greater efficiency, attention focused on the need to manage not just the buildings but the entire resources used by organisations in the generation of their wealth – facilities management (FM). It is not a new concept but one that has progressed from being used by a handful of enlightened companies to become the fastest-growing property and resource management sector in our industry.

FM seeks to create a framework that embraces the traditional estate management functions of property maintenance, lighting and heating with increasingly analytical reviews of space occupation/planning, asset registers, health and safety registers, and activity flow throughout the premises. Hence the term facilities is used to include all the buildings, furnishings, equipment and environment available to the workforce while pursuing the company's business goals.

The success of FM has been greatly enhanced by the development of reliable and powerful computer technology together with the boom in personal computers that has made serious data handling affordable to all. The use of databases to control the occupational activities of buildings is both reactive and proactive, with the latter gaining in importance. The reactive use allows data on the performance of the workplace to be collected and stored, these in turn are available for historical analysis that can be used proactively to identify recurring trends and anticipate operational problems, so eliminating waste.

Every FM application in industry and commerce is in effect a one-off system; it must address the priorities of the company but is actually assembled from a series of independent modules that operate from a universal FM platform. The emerging industry standard platform is based on the computer-aided design technology used extensively in the design of buildings; this has been developed into powerful computer-aided facilities management (CAFM) systems. CAFM systems are increasingly likely to serve as an indispensable source of reference for the project manager and project team in drawing up the project brief for buildings of similar function. The pairing of facilities management and project management in this way should enable the procurement of increasingly efficient property.

The CIOB has become aware of the vast array of bespoke, tailor-made Facilities Management Contracts which are prevalent in the FM industry. Many of these contracts are often based upon models from other industries, and suffer from a lack of focus on FM contractual issues.

As part of ongoing strategy in FM, the CIOB in partnership with Cameron McKenna published the first standard form of Facilities Management Contract in 1999 as the industry's first *Standard Form of Facilities Management Contract*. This first edition contract was designed with flexibility so that it can be used for all categories of private and public sector works, where facilities are provided or managed by the service provider or facilities manager. This contract has proved extremely popular, selling out two print runs and establishing itself as the benchmark *Facilities Management Contract* for the industry.

The CIOB in association with CMS Cameron McKenna published the second edition of its *Standard Form of Facilities Management Contract* in 2001. The second edition of the contract was built on the success of the first, being updated and revised to deal specifically with areas where it was felt the first edition warranted revision.

APPENDIX 20 Value for money project framework

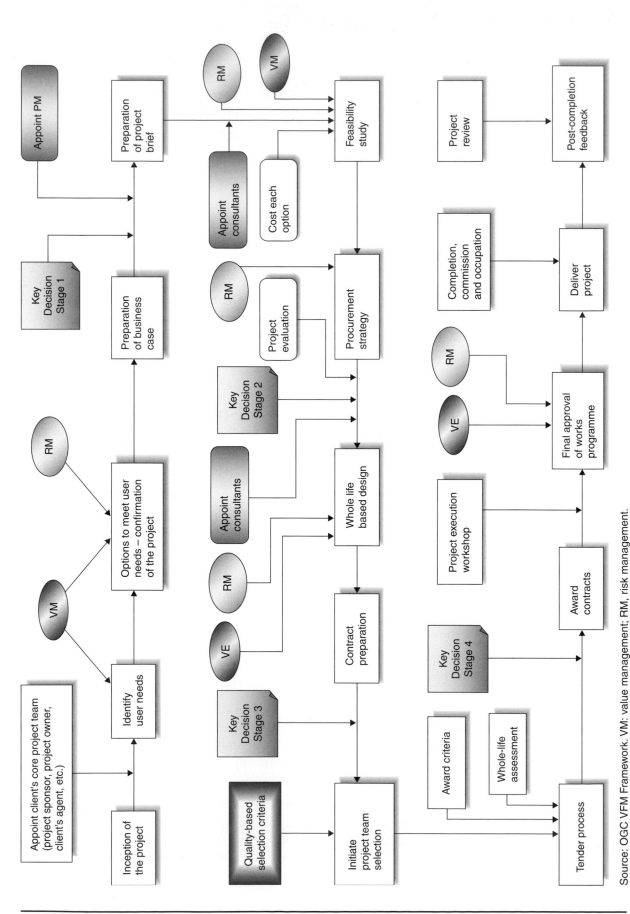

Source: OGC VFM Framework. VM: value management; RM, risk management.

Part 2
Project handbook

Project handbook

Introduction

The purpose of the handbook is to guide the project team in the performance of its duties, which are the design, construction and completion of a project to the required specifications within the approved parameters of the contract budget and to schedule.

In practice a project handbook should be concise, clear and consistent with all other contract documentation and terms of engagement. The emphasis should be to identify policies, strategies and the lines of communication and key interfaces between the various parties. It is important that the handbook is tailored to fit the needs of each project. The comprehensive format given here would be too bulky for some projects, with the danger of it being ignored.

The handbook is prepared by the project manager in consultation with the project team where possible at the beginning of the pre-construction stage and describes the general procedures to be adopted by the client and the team. It comprises a set of ground rules for the project team. It differs from the Project Execution Plan, which is primarily written for the client and its funding partners giving a route map through the stages and processes of the project demonstrating financial control and a modus operandum to achieve the project objectives.

The handbook is not a static document and it is anticipated that changes and amendments will be required in accordance with procedures as later outlined. Consequently a loose-leaf format should be adopted to facilitate its updating by the project manager who is the only person authorised to co-ordinate and implement revisions. Copies of the handbook will be provided to each nominated member of the project team as listed under Parties to the project.

Aims of the handbook

Its aims are to identify responsibilities and co-ordinate the various actions and procedures from other documents/data already or currently or likely to be prepared into one authoritative document covering as a rule, and depending on the nature/scope of the project, the main elements and activities outlined in the following sections.

Parties to the project

This section will include the following items:

- A list of all parties involved in the project including those employed by the client and their contact details (addresses, phone and fax numbers and e-mail).

- The name of the project manager responsible for the project together with details of his duties, responsibilities and authority (see Appendix 1 of Part 1).

■ Details of other team members and/or stakeholders involved, complete with their duties, responsibilities and contact details.

Organisation charts indicating line and functional relationships, contractual and communication links and any changes to suit the various stages/phases of the project.

Third parties

This section will provide the names and contact details of all legal authority departments, public utilities, hospitals, doctors, police stations, fire brigade, trade associations, adjoining landowners, adjacent tenants and any other bodies or people likely to be involved.

Roles and duties of the project team

The information provided should be the minimum necessary to facilitate the understanding of the roles of the others involved by each member of the team. The services to be provided are described by reference to standard agreements/contracts with any amendments and additions included. The aim is to ensure that there are no gaps or overlaps.

Project site

Details will be provided of prevailing relevance of arrangements for demolition, clearing and diversion of existing services, hoardings and protection to adjacent areas (e.g. noise pollution).

General administration including communication and document control

The project manager will be responsible for the following items:

■ The adequacy of all aspects of project resourcing (staffing, equipment and aids, site offices and welfare accommodation).

■ Office operating systems and routines so that the staff know them and they are applied consistently and efficiently.

■ Providing suitable working accommodation and facilities for members of the project team and for meetings/group discussions.

Action will need to be taken by the project manager in respect of documentation control, storage, location and retrieval; this will affect:

■ Letters, contract documents, reports, drawings, specifications, schedules, including financial and all specialist fields (e.g. facilities management, technology, health/safety/environmental).

■ Accessibility for updating.

■ Records for all documents/files and control of their movement.

■ Office security: (1) storage of legal documents (originals and duplicates); (2) entry safeguard, fire and intruder alarms.

■ Retention of documents/files on project completion/suspension: (1) archive storage – legal and contractual time limits; (2) dead files – removal/destruction and their register.

All correspondence should be headed by the project title and identified by:

- subject/reference of communication
- addressee's full details
- those parties receiving copies.

Each piece of correspondence should refer to a single matter or a series of direct and closely linked matters only.

Distribution of copies should be decided on the basis of the subject matter and confidentiality against a predetermined list of recipients.

All communications between the parties of the project involving instructions must be given in writing and the recipients should also confirm it in the same manner.

Contract administration

Contract conditions

It is essential that there is an understanding of the terms of all contracts and their interpretation by all concerned. The role of parties, their contribution and responsibilities, including relevance of time-scales and client–project manager operating and approvals pattern will have to be established.

Contract management and procedures

Matters associated with contract management will include forms of contract for contractors/subcontractors; works carried out under separate direct contract; procedures for selection and appointment of contractors/subcontractors; checklists for design team members and consultants meeting their supervisory and contractual obligations (e.g. inspections and certification); the placement of orders for long delivery components and the preparation of contract documents.

Tender documentation

Design and specifications details to be included; tender analysis and reporting; lists of tenderers and interview procedures; system for the preparation of documents and their check; award and signature arrangements.

Assessment and management of variations

(See Appendix 13 of Part 1 for change management checklist and Appendix 16 for dispute resolution processes.)

Extensions of time

- The project manager has responsibility for ensuring that there is early warning, hence creating the possibility of alternative action/methods to prevent delay and additional costs.
- A schedule should be prepared, stating the grounds for extension, relevant contract clauses and forecast of likely delay and cost.
- The involvement and possible contribution to the solution of problems of other parties affected should be established.

■ A procedure will need to be available for extension approval. If relevant, the disputes procedure may be invoked.

Loss and expense

■ Applicable procedures are covered under standard or in-house forms of contract relevant to the specific project.

Indemnities, insurances and warranties

Relevant provisions depend on the nature of the project. However they are usually governed by the conditions specified by the forms of 'model' contracts/agreements issued by professional bodies or those in common use in the construction industry.

Typical examples of insurances applicable to construction projects include:

■ Contractors' all-risk (CAR) policies, usually covering loss or damage to the works and the materials for incorporation in the works; the contractor's plant and equipments including temporary site accommodation; the contractor's personal property and that of his employees (e.g. tools and equipment). The CAR policy is normally taken out by the contractor but should insure in the joint names of the contractor and the client (employer). The subcontractors may or may not be jointly insured under the CAR policy.

■ Public liability policy – this insures the contractor against the legal liability to pay damages or compensation or other costs to anyone who suffers death, bodily injury or other loss or damage to their property by the activities of the contractor.

■ Employers' liability policy – every contractor will have this either on a company-wide basis, covering both staff and labour, or on a separate basis for the head office and for each site separately.

■ Professional indemnity (PI) – the purpose of this is to cover the liabilities arising out of 'duty of care'. Typically the consultants (including the project manager) will require this policy to cover their design or similar liabilities and liabilities for negligence in undertaking supervision duties. In the case of a design and build contract, the contractor has to take out a separate PI policy, as designing is not covered by the normal CAR policy.

Design co-ordination

The project brief will be reviewed jointly by client and project manager with the aim of confirming that all relevant issues have been considered. These may include:

■ health and safety obligations

■ environmental requirements

■ loading considerations

■ space and special accommodation requirements ⎫

■ standards and schedule of finishes   ⎬ Eventual user's need

■ site investigation information/data

■ availability of necessary surveys and reports

■ planning consent and statutory approvals

■ details of internal and external constraints.

Part 2 project handbook

The project manager will need to seek the client's approval to issue the brief and relevant information to the design team and other consultants. Among the other duties which fall to the project manager are the following:

- Defining the roles and duties of the project team members.

- Responsibility for the drawings and specifications:
  - establishing format (e.g. CAD compatibility issues), sizes and distribution and seeking comments upon their content and their timing
  - the issue of tender drawings and specifications
  - advising contractors and subcontractors of the implications of the design
  - setting requirements for: (1) shop/fabrication drawings; (2) test data; (3) sample and mock-ups.

- Monitoring the production of the outline proposals for the project by:
  - reviewing sketch plans and outline specifications in terms of the brief
  - preparing the capital budget and reconciling this with the outline budget
  - appraising the implications of the schedule
  - effecting reconciliation with the project master schedule
  - finalising the outline proposals making recommendations/presentations to the client and seeking the latter's approval to proceed.

- Monitoring design work at the pre-tender stage by:
  - reviewing with the consultants concerned the client's requirements, brief documentation and their sectional implications
  - agreeing team members' input and identification of items needing client's clarification
  - reviewing with the client any discrepancies, omissions and misunderstandings, seeking their resolution and confirming to the team.

- Agreement of overall design schedule and related controls.

- Identification of items for pre-ordering and long delivery preparation of tender documents, client's approvals and placement of orders and their confirmation.

- Monitoring production of drawings and specifications throughout the various stages of the project and their release to parties concerned.

- Arranging presentations to the client at appropriate stages of design development and securing final approval of tender design.

Change management

- Reviewing with the design team and other consultants any necessary modifications to the design schedule and information required schedules (IRS) in the light of the appointed contractors'/subcontractors' requirements and reissuing revised schedule/IRS.

- Preparing detailed and specialist designs and subcontract packages including bills of quantities.

- Making provision for adequate, safe and orderly storage of all drawings, specifications and schedules including the setting up of an effective register/records and retrieval system.

LIVERPOOL JOHN MOORES UNIVERSITY
LEARNING & INFORMATION SERVICES

The project manager must ensure that the client is fully aware that supplementary decisions must be obtained as the design stages progress and well within the specified (latest) dates in order to avoid additional costs. Designs and specifications meeting the client's brief and requirements are appraised by the quantity surveyor for costs and are confirmed to be within the budgetary provisions.

Handling changes will require a series of actions. The project manager will be responsible for these activities:

■ Administering all requests through the change order system (see Table 5.3 and Appendix 17 for checklist and specimen form).

■ Retaining all relevant documentation.

■ Producing a schedule of approved and pending orders which will be issued monthly.

■ Ensuring that no changes are acted upon unless decisioned.

■ Considering amendments and alterations to the schedules and drawings within the provisions of the applicable contract/agreement.

■ Initial assessment of any itemised request for change made by the client taking due account of the effect on time.

Action by consultants in relation to variations will include the following items:

■ Securing required statutory/planning approvals and cost-checking revised proposals. Confirmation of action taken to project manager.

■ Design process and preparation of instructions to contractors involved.

■ Cost agreement procedure for omissions and additions, i.e. estimates, disruptive costs, negotiations and time implications.

Site instructions

Site instructions must be issued in writing and confirmed in a similar manner by recipients.

Site instructions which constitute variations can be categorised as:

■ normal

■ special (e.g. concerned with immediate implementation as essential for safety, health and environmental protection aspects)

■ extension of time required or predicted

■ additional payments involved or their estimate.

Site instructions will be binding if they are issued and approved in accordance with the contract provisions.

Cost control and reporting

The quantity surveyor has overall responsibility for cost monitoring and reporting with the assistance of and input from the design team, other consultants and contractors.

Part 2 project handbook

Action at the pre-construction stage involves the following items:

- The preparation of preliminary comparison budget estimates.

- The agreement of the control budget with the project manager.

- Project budget being prepared in elemental form; the influence of grants is identified.

- The establishment of work packages and their cost budgets.

- Costing of change orders.

Other elements associated with work control are as follows:

- Assessment of cost implications for all designs, including cost comparison of alternative design solution.

- Value analysis procedures, including cost in use.

- Comparison of alternative forms of construction using data on their methodology and costs.

- Comparison of cost budgets and tenderers' prices at subcontract tender assessment.

- Tenders which are outside the budget and which require an input from the project manager on such matters as:

  - alteration of specifications to reduce costs

  - acceptance of tender figure and accommodating increased cost from contingency; alternatively the client may accept the increase and seek savings from other areas

  - possible retendering by alternative contractors.

- Production of monthly cost reports including:

  - variations since last report incorporating reasons for costs increase/decrease

  - current projected total cost for the project

  - cash flow for the project: (1) forecast of expenditure; (2) actual cash flow as schedule monitoring device indicating potential overspending and any areas of delay or likely problems.

The report should be agreed with and issued to the project manager who will:

- give advice and initiate action on any problems that are identified

- arrange distribution of copies according to a predetermined list.

Planning, scheduling and progress reporting

Planning is a key area and can have a significant effect on the outcome of a project. The handbook will set out the composition and duties of the planning support team and the appropriate techniques to be used (e.g. bar charts, networks). The planning and scheduling will then follow the steps set out in Appendix 3:

- Preparation of an outline project schedule, which will include co-ordination of design team contractors' and client's activities then seeking of the client's acceptance.

- Production of an outline construction schedule indicating likely project duration and the basis for determining the procurement schedule.

- Production of an outline procurement schedule including the latest date for placement of orders (materials equipment contractors) and design release dates.

- Modifications if necessary to the outline construction schedule due to constraints.

- Production of the outline design schedule including necessary modifications due to external limitations.

- Preparation of the project master schedule.

- Preparation of a short-term schedule for the pre-construction stage; this will be reviewed monthly.

- Production of a detailed design schedule in consultation with and incorporating design elements from the design team members concerned including:

  o scheme design schedule

  o drawing control schedule

  o client decision schedule

  o agreement by client consultants and project manager.

- Reviewing the outline procurement schedule and its translation into one, which is detailed.

- Preparation of a works package schedule.

- Production of schedules for bills of quantities procurement, including identification of construction phases for tender documentation and production of tender documentation control.

- Expansion of the outline construction schedule into one, which is detailed.

- Preparation of schedules for:

  o enabling works

  o fitting out (if part of the project)

  o completion and handover

  o occupation/migration (if part of the project).

Progress monitoring and reporting procedures should be on a monthly basis and agreed following consultation with consultants and contractors. Reports will need to be supplied to the project manager who will report to the client.

Meetings

Meetings are required to maintain effective communications between the project manager, project team and the other parties concerned, e.g. those responsible for industrial relations and emergencies as well as the client.

The frequency and location of meetings and those taking part will be the responsibility of the project manager. Meetings held too frequently can lead to a waste of time whereas communications can suffer where meetings are infrequent.

Appendix A contains details of typical meetings and their objectives.

Procedures for meetings include:

■ agenda – issued in advance stating action/submissions required

■ minutes and circulation list (time limits involved)

■ written confirmation and acknowledgement of instructions given at meetings (time limit involved)

■ reports/materials tabled at meetings to be sent in advance to the chair.

Selection and appointment of contractors

The project manager as the client's representative has the responsibility with the support of relevant consultants for the selection and appointment of:

■ contractors, e.g. main, management, design and build

■ contractors, e.g. specialist, works, trade.

The various processes associated with this activity are summarised below.

■ Selection panel appointments relevant to the nature and scope of tender to be awarded. Nomination of a co-ordinator (contact) for all matters concerned with the tender.

■ Establishment of selection/appointment procedures for each stage.

Pre-tender

Pre-tender activities will include the following:

■ Assessment of essential criteria/expertise required for a specific tender.

■ Preparation of long (provisional) list embracing known and prospective tenderers.

■ Checks against database available to project manager, especially financial viability and quality of past and current work; possible use of telephone questionnaire to obtain additional data.

■ Potential tenderers invited to complete/submit selection questionnaire; short list finalised accordingly.

■ Arrangements for pre-qualification interview including prior issue of the following documentation relevant to the project to the prospective tenderer, with interview agenda outline of special requirements and expected attendees to cover:

　○ general scope of contract works and summary of conditions

　○ preliminary drawings and specifications

　○ summary of project master and construction schedules

　○ pricing schedule

　○ safety, health and environmental protection statement

　○ labour relations statement

　○ quality management outline.

■ Tender and reserve lists finalised.

Tendering process

The tendering process includes the following activities:

- Selected tenderers confirm willingness to submit bona fide tenders. Reserve list is employed in the event of any withdrawals and selection made in accordance with placement order.

- Tender documents are issued and consideration is given by both parties to whether mid-tender interview is required or would be beneficial.

- Interview is arranged and agenda issued.

On receipt of all tenders carry out:

- evaluation of received tenders

- arrangements for post-tender interview and prior issue of agenda

- final evaluation and report

- pre-order check and approval to place order.

Safety, health and environmental protection

The handbook should draw attention to the specific and onerous duties of the client and other project team members under the CDM Regulations and include procedures to ensure they cannot be overlooked.

It is the responsibility of the principal contractor to formulate the health and safety plan for the project to be adhered to by all contractors in accordance with the CDM Regulations and taking account of other applicable legislation.

Contractors are required as part of their tender submission to provide copies of their safety policy statement which outlines safe working methods that conform to the CDM Regulations.

Other matters which come within the remit of the principal contractor are as follows:

- The establishment and enforcement within the contractual provisions of rules, regulations and practices to prevent accidents, incidents or events resulting in injury or fatality to any person on the site or damage or destruction to property, equipment and materials of the site or neighbouring owners/occupiers.

- Arranging first-aid facilities, warning signals and possible evacuation as well as the display of relevant notices, posters and instructions.

- Instituting procedures for:
  - regular inspections and spot checks
  - reporting to the project manager (with copies to any consultants concerned) on any non-compliance and the corrective or preventive action taken
  - hazardous situations necessitating work stoppage and in extreme cases closedown of the site.

Quality assurance (QA) – outline

This is applicable only if QA is operated as part of contractual provisions.

It is critical for the client to understand the operation of a QA scheme, its application and limits of assurance and the need for defects insurance.

Procedures and controls will need to be established to ascertain compliance with design and specifications and to confirm that standards of work and materials quality have been attained.

The consultants will review details of their quality control with the project manager.

The contractors' quality plan will indicate how the quality process is to be managed, including control arrangements for subcontractors.

Responsibility for monitoring site operation of QA administration and control procedures for the relevant documents will need to be established.

As an alternative to QA any procedures for the management of quality should be included in the handbook (see Quality Management in Chapter 4).

Disputes

Procedures for all parties involved in the project in the event of disagreement and disputes are to be specified in accordance with the contractual conditions/provisions which are applicable.

Signing off

Any procedures for signing-off documents should be specified. Signing-off points may occur progressively during stages of the project and be incorporated in a 'milestone schedule'. Details should include permitted signatories and a distribution list.

Reporting

The following reports are examples of what might be prepared.

Project manager's progress report

To be issued monthly and include details of:

- project status
    - updated capital budget
    - accommodation schedule
    - authorised change orders during the month
    - other relevant matters
- operational brief
- design development status
- cost plan status and summary of financial report

- schedule and progress:
  - design
  - construction
- change/variation orders
- client decisions and information requirements
- legal and estates
- facilities management
- fitting out and occupation/(migration) planning
- risks and uncertainties
- update of anticipated final completion date
- distribution list.

Consultant's report

Issued monthly and including input from consultants and containing details on:

- design development status
- status of tender documents
- information produced during the month
- change orders/design progress
- information requirements/requests status
- status of contractor/subcontractor drawings/submittals
- quality control
- distribution list.

Financial control (QS) report

Issued monthly and including:

- reconciliation capital sanction/capital budget
- updated cost plan and anticipated final cost projection
- authorised change orders – effects
- pending change orders – implications
- contingency sum
- cash flow
- VAT
- distribution list.

Part 2 project handbook

Daily/weekly diary

Prepared by each senior member of the project team and filed in its own separate loose-leaf binder for quick reference and convenient follow-up. Diaries are made accessible to the project manager and typically contain:

- a summary of forward and ad hoc meetings and persons attending
- a summary of critical telephone conversations/messages
- documents received or issued
- problems, comments or special situations and their resolution
- schedule status (e.g. work package progress or delays)
- critical events and work observations
- critical instructions given or required
- requests for decisions or actions to be taken
- an approximate time of day for each entry
- a distribution list.

Construction stage

The handbook will include procedures for the following activities:

- Issuing drawings specifications and relevant certificates to contractors.
- Actioning the consultants' instructions, lists, schedules and valuations.
- Aspects prior to commencement such as:
  - recording existing site conditions including adjacent properties
  - ensuring that all relevant contracts are in place and that all applicable conditions have been met
  - confirming that all risk insurance for site and adjacent properties is in force
  - ensuring that all site facilities are to the required standard including provisions for safety health and environmental protection.
- Control of construction work including:
  - reviewing a contractor's preliminary schedule against the master schedule and agreeing adjustments
  - ensuring checks by the main contractor on subcontractor schedules
  - checking and monitoring for all contractors the adequacy of their planned and actual resources to achieve the schedule
  - approvals for subletting in accordance with contractual provisions
  - reporting on and adjusting schedules as appropriate
  - checks for early identification of actual or potential problems (seeking client's agreement to solutions of significant problems).
- Controls for variations and changes (see Appendix 13 of Part 1).
- Controls for the preparation and issue of change orders (see Figure 5.1 and Appendix 13 of Part 1).

- Processing the following applications for client's action:

  ○ interim payments from consultants and contractors

  ○ final accounts from consultants

  ○ final accounts from contractors subject to receipt of relevant certification

  ○ payment of other invoices.

- Making contact and keeping informed the various authorities concerned to facilitate final approvals.

- The design team and other relevant consultants to supervise and inspect works in accordance with contractual provisions/conditions and participate in and contribute to:

  ○ the monitoring and adjustment of the master schedule

  ○ controls for variations and claims

  ○ identification and solutions of actual or potential problems

  ○ subletting approvals

  ○ preparation of change orders.

Operating and maintenance (O&M)

The procedures for fitting out should be designed to avoid divided responsibility in the case of failure of parts of the building or its services systems. The procedures to be used in the handbook can be developed by reference to the relevant sections in Chapters 5–7. They should include adequate arrangements for the management of any interfaces between contracts or work packages. It is especially vital to have procedures for:

- the transition of commissioning data record drawings and O&M manuals from one contract to the next

- confirmation that all relevant handover documentation and certification has been completed.

Engineering services commissioning

Engineering services commissioning is part of the construction stage. It is the main contractor's responsibility which is delegated to the services subcontractors. Action is taken in two stages: pre-construction and construction/post-construction.

Pre-construction

- Ensuring the client recognises engineering services commissioning as a distinct phase of the construction process starting at the strategy stage.

- Ensuring that the consultants identify all services to be commissioned and defining the responsibility split for commissioning between designer, contractor, manufacturer and client.

- Identifying statutory and insurance approvals required and planning to meet requirements and obtain approvals.

- Co-ordinating the consultants' and client's involvement in commissioning to ensure conformity with the contract arrangements.

■ Arranging single-point responsibility for control and the client's role in the commissioning of services.

■ Ensuring contract documents make provision for services commissioning.

Construction and post-construction

■ Ensuring relevant integration within construction schedules.

■ Monitoring and reporting progress and arranging corrective action.

■ Ensuring provision and proper maintenance of records, test results, certificates, checklists, software and drawings.

■ Arranging for or advising on maintenance staff training post-contract operation and specialist servicing contracts.

Examples of a checklist and documentation are given in Appendices C and D.

Completion and handover

The closely interlinked processes of completion and handover are very much a hands-on operation for the project manager and his team. This stage provides the widest and closest involvement with the client. Completion and handover require careful attention because they determine whether or not the client views the whole job as successful.

Completion

Handbook procedures may cover two sorts of agreement:

■ Agreements for partial possession and phased (sectional) completion (if required):

○ Access inspections, defects, continuation of other works and/or operation of any plant/services installation material, obstructions or restrictions.

○ Certification on possession of each phase; responsibility for insurance.

■ Agreements and procedures associated with practical completion:

○ User/tenant responsibility for whole of the insurance.

○ Provision within a specified time limit of complete sets of as-built and installed drawings, M&E and other relevant installations/services data as well as all operating manuals and commissioning reports.

○ Storage of equipment/materials except those required for making good any defects.

○ Access for completion of minor construction works, rectification of defects, testing of services, verification of users' works and other welfare and general facilities.

Appendix 18 of Part 1 provides a typical checklist at the practical completion stage.

Part 2 project handbook

Handover

Procedures are needed for the following activities:

- To ensure that handover only takes place when all statutory inspections and approvals have been satisfactorily completed and subsequently to arrange that all outstanding works and defects are resolved before expiry of the defects liability period.

- To provide and agree a countdown schedule with the project team (examples of handover inspections and certificate checklists are given in Appendix E).

- To define responsibilities for all inspections and certificates.

- To monitor and control handover countdown against the schedule.

- To control pre-handover arrangements if the client has access to the building before handover.

- To deal with contractors who fail to execute outstanding works or correct defects including the possibility of implementing any contra-charging measures available under the contract. Agree and set up a procedure for contra-charging.

- To monitor and control any post-handover works which do not form part of the main contract.

- To monitor and control outstanding post-completion work and resolution of defects which form part of the main contract.

- To manage the end of the defects liability period and implement relevant procedures.

- To establish arrangements for the final account, issuing the final certificate and carrying out the post-completion review/project evaluation report.

Client commissioning and occupation

Client commissioning

Client commissioning will involve the following handbook procedures:

- Arranging the appointment of the commissioning team in liaison with the client and establishing objectives (time, cost and specifications) and responsibilities at the feasibility and strategy stages.

- Preparation of a comprehensive commissioning and equipment schedule.

- Arranging access to the works for the commissioning team and client personnel during construction, including observation of engineering services commissioning.

- Ensuring co-ordination and liaison with the construction processes and consultants.

- Preparing new work practice manuals and, in close liaison with the client's/user's facilities management team, arranging staff training and recruitment/secondment of additional staff (e.g. aftercare engineer to support the client during the initial period of occupancy).

- Deciding the format of commissioning test and calibration records.

- Renting equipment to meet short-term demands.

- Deciding quality standards.

- Monitoring and controlling commissioning progress and reporting to the client.

- Reviewing post-contract the operation of the building at 6, 9 and 12 months: improvements, defects, corrections and related feedback.

Appendix F contains the relevant checklist.

Occupation

Occupation can be part of the overall project or a separate project on its own. A decision to this effect is made at the strategy stage with the client or user.

The separate stages of occupation are set out below. Figures 7.1–7.4 illustrate these procedures graphically and Appendix F provides an example of an occupation implementation plan.

Structure for implementation In order to achieve the necessary direction and consultations, individuals and groups are appointed, e.g.

- project executive (client/occupier/tenant)

- occupation co-ordinator (project manager)

- occupation steering group:
    - chairman co-ordinator and functional representatives
    - concerned with overall direction for:
    - construction schedule
    - technology
    - space planning
    - facilities for removal
    - user representation
    - costs and budget outline

- senior representative meeting:
    - chair (functional representative on steering group) co-ordinator and senior representatives of majority of employees
    - concerned with consultations on:
    - space planning
    - corporate communications
    - construction schedule problems
    - technology

- local representative groups:
    - chaired by manager/supervisor of own group
    - concerned with consultation at locations and/or departmental levels in order to ensure procedures for regular communications.

Scope and objectives (regularly reviewed)

- identification of who is to move (project executive)
- agreement on placement of people in new locations (steering group)
- decision on organisation of move (steering group):
  - all at once
  - several moves
  - gradual flow
- reviewing time constraints (steering group):
  - construction
  - commercial
  - holidays
- identification of risk areas, e.g.
  - construction delays and move flexibility
  - organisational changes
  - access problems
  - information technology requirements
  - furniture deliveries and refurbishment
  - retrofit requirements.

Methodology

- listing special activities needed to complete the move, e.g.
  - additional building work
  - communications during move
  - provision of necessary services and move support
  - corporate communications
  - removal administration
  - furniture procurement
  - removal responsibility in each location/department
  - financial controls
  - access planning
- preparation of task list for each special activity, confirmation of person responsible and setting the schedule of project meetings
- production of outline and subsequently detailed schedule.

Organisation and control

- steering group establishes 'move group' to oversee the physical move
- production of 'countdown' schedule (move group)
- identification of external resources needed (move group), e.g.
  - special management skills
  - one-off support tasks
  - duplication of functions during move

■ reporting to client external support needs and costs (steering group)

■ preparation of monitoring and regular review of actual budget (steering group), e.g.

- ○ dual occupancy

- ○ special facilities

- ○ additional engineering and technology needs

- ○ planning and co-ordinating process

- ○ inflation

- ○ external resources

- ○ non-recoverable VAT

- ○ contingencies.

PART 2 APPENDICES

APPENDIX A Typical meetings and their objectives

Steering group/team
- to consider project brief, design concepts, capital budget and schedules
- to approve changes to project brief
- to review project strategies and overall progress towards achieving client's goals
- to approve appointments for consultants and contractors.

Project team
- to agree cost plan and report on actual expenditure against agreed plan
- to review tender lists and tenders received and decide on awarding work
- to report on progress on design and construction schedules
- to review and make recommendations for proposed changes to design and costs, including client changes; to approve relevant modifications to project programmes.

Design team
- to review, report on and implement all matters related to design and cost
- to determine/review client decisions
- to prepare information/report/advice to project team on (1) appointment of sub/specialist contractors; (2) proposed design and/or cost changes
- to review receipt, co-ordination and processing of subcontractors' design information
- to ensure overall co-ordination of design and design information.

Finance group/ team
- to review, monitor and report financial, contractual and procurement aspects to appropriate parties
- to prepare a project cost plan for approval by the client
- to prepare and review regular cost reports and cash flows, including forecasts of additional expenditure
- to review taxation matters
- to monitor the preparation and issue of all tender and contract documentation
- to review cost implications of proposed client and design team changes.

Project team (programme/ progress meeting)
- to provide effective communication between teams responsible for the various phases of the project
- to monitor progress and report on developments, proposed changes and schedule implications
- to review progress against schedules for each stage/section of the project/ works and identify any problems
- to review procurement status
- to review status of information for construction and contractors' sub-contractors' requests for information.

**Project team
(site meeting)**

Main contractor report tabled monthly to include details on:

- quality control
- progress
- welfare (health, safety, canteen, industrial relations)
- subcontractors
- design and procurement
- information required
- site security
- drawing registers.

Reports/reviews (including matters arising at previous meetings) from:

- architect
- building services
- facilities management
- information technology
- quantity surveyor.

Statutory undertakings and utilities:

- telephones
- gas
- water
- electricity
- drainage.

Approvals and consents:

- planning
- Building Regulations
- local authority engineer
- public health department
- others.

Information:

- issued by design team (architect's instructions issued and architect's tender activity summary)
- required from design team
- required from contractor.

APPENDIX B Selection and appointment of contractors
B1 Pre-tender process

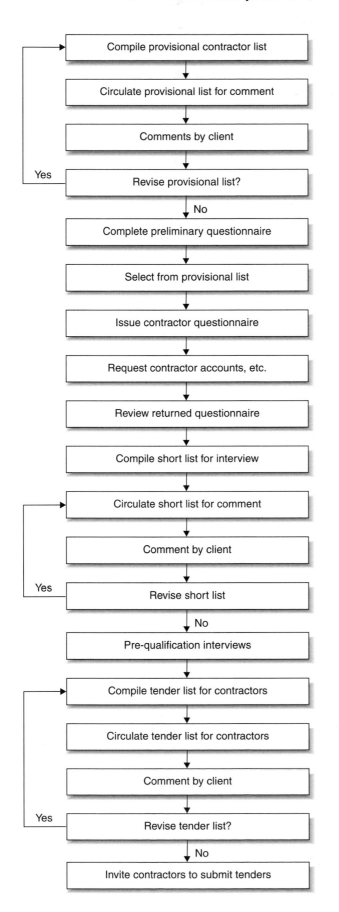

Checked by | Comments

APPENDIX B   Selection and appointment of contractors
B2 Initial questionnaire

Project Management Ltd		Form Q1
Ref. number:		
Contract title:		
Item no.	Question	Response
1.0	Turnover of company?	
2.0	What is the value of contracts secured to date?	
3.0	What is the largest current contract?	
4.0	Is the contractor willing to submit a tender?	
5.0	Is the contractor willing to work with all team members?	
6.0	Is the contract period acceptable?	
7.0	If not how long to complete works?	
8.0	Is the anticipated tender period acceptable?	
9.0	If not how long to tender?	
10.0	What are the mobilisation periods of (a) completion of drawings? (b) fabrication? (c) start on site from order?	
11.0	Is the labour used direct self-employed or subcontract?	
12.0	What element of the contract will be sublet?	
Comments Signature and date _____		

APPENDIX B Selection and appointment of contractors
B3 Selection questionnaire

1 Name of company:	
2 Address:	
3 Telephone no.:	Facsimile no.:
4 Nature of business:	

5 Indicate whether
   (a) manufacturer
   (b) supplier
   (c) subcontractor
   (d) main contractor
   (e) design and build contractor
   (f) management contractor

6 Indicate whether
   (a) sole trader
   (b) partnership
   (c) private
   (d) public

7 Company registration number:	
8 Year of registration:	
9 Bank and branch:	
10 VAT registration number:	
11 Tax exemption certificate number:	Date of expiry:
12 State annual turnover of current and previous four years:	
13 State value of future secured work:	
14 State maximum and minimum value of works undertaken:	
15 Are you registered under BS 5750/ISO 9000?:	
16 State previous projects undertaken with this company	
17 Are you prepared to sign a design warranty?	
18 Are you prepared to provide a performance bond?	
19 Are you prepared to provide a parent company guarantee?	
20 Do you operate a holiday with pay scheme?	
21 State when stamps last purchased:	
22 Do you contribute to the CITB?	

APPENDIX B   Selection and appointment of contractors
B3 Selection questionnaire (*cont'd*)

---

23 Do you have a safety policy?

---

24 Are you competent and willing to act as principal contractor under CDM?

---

25 Employer's liability insurance:

Insurer:

Policy no.:

Expiry date:

Limit of indemnity:

---

26 Third party insurance:

Insurer:

Policy no.:

Expiry date:

Limit of indemnity:

---

27 Which elements do you sublet?

---

28 List of projects of similar size and complexity:

Project 1.:

Address:

Architect:

Contact: Telephone no.:

Contractor:

Contact: Telephone no.:

Value:

Year completed:

Project 2.:

Address:

Architect:

Contact: Telephone no.:

Contractor:

Contact: Telephone no.:

Value:

Year completed:

Project 3.:

Address:

Architect:

Contact: Telephone no.:

Contractor:

Contact: Telephone no.:

Value:

Year completed:

APPENDIX B   Selection and appointment of contractors
B4 Pre-qualification interview agenda

Project Management Ltd	Form A1

Ref. number:

Contract title:

1.0	**Introduction**
1.1	Purpose of meeting
1.2	Introduction to those present
2.0	**Description of overall project and schedule**
2.1	General description of the project
2.2	Master schedule in summary
2.3	General description of contract
3.0	**Explanation of contract terms and conditions**
3.1	Outline and scope of contract
3.2	Responsibilities of the contractor
3.3	Outline of contract conditions including any significant amendments
3.4	Schedule
3.5	Specification
3.6	Drawings
3.7	Preliminaries
3.8	Budget prices
4.0	**Project organisation**
4.1	Site administration and project team
4.2	Setting out and dimensional control
4.3	Materials handling and control
4.4	Site establishment
4.5	Contractor supervision and on-site representative
4.6	Labour relations
4.7	Quality management
4.8	Health and safety plan
5.0	**Tendering**
5.1	Period of tendering
5.2	Mid-tender interview
5.3	Tender return date address and contact name
6.0	**Actions required**
6.1	Summary of actions and date deadlines

APPENDIX B   Selection and appointment of contractors
B5 Tendering process checklist

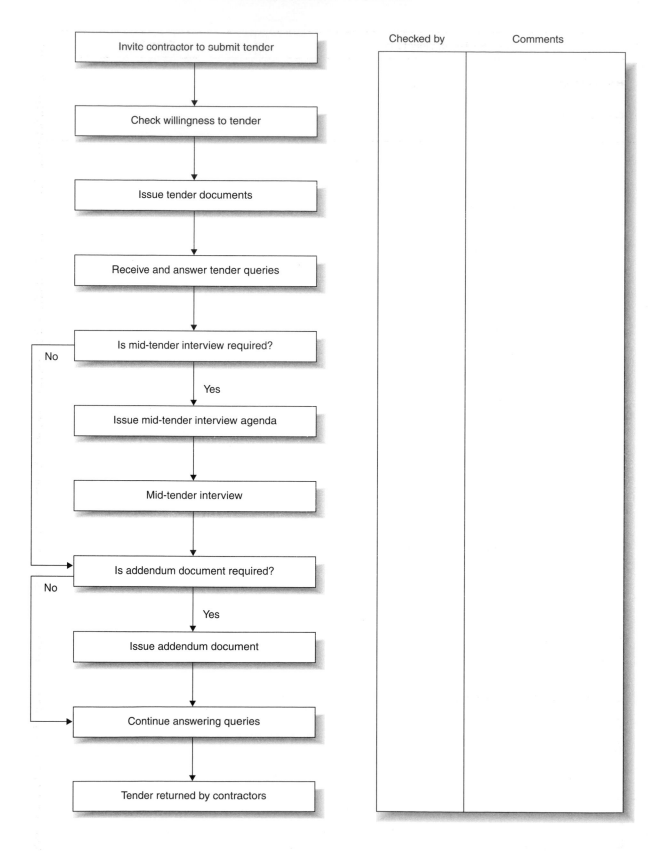

Checked by   Comments

Invite contractor to submit tender

Check willingness to tender

Issue tender documents

Receive and answer tender queries

Is mid-tender interview required?   No

Yes

Issue mid-tender interview agenda

Mid-tender interview

Is addendum document required?   No

Yes

Issue addendum document

Continue answering queries

Tender returned by contractors

APPENDIX B   Selection and appointment of contractors
B6 Tender document checklist

Project Management Ltd	Form C1

Ref. number:

Contract title:

☐   Invitation to tender

☐   Introduction and scope of contract

☐   Instructions to tenderers

☐   Form of tender

☐   General preliminaries

☐   Particular preliminaries

☐   Form of contract and amendments

☐   Contract schedule

☐   Method statement

☐   Quality management

☐   Project health and safety plan

☐   Project labour relations

☐   Specification

☐   List of drawings

☐   Bill of quantities or pricing schedule

☐   General summary

☐   Declaration of non-collusion

☐   Performance bond

☐   Warranty

☐   Soil report

☐   Contamination reports

Other documents (please list below)

☐

☐

☐

☐

☐

☐

☐

APPENDIX B   Selection and appointment of contractors
B7 Mid-tender interview agenda

Project Management Ltd	Form A2

Ref. number:

Contract title:

1.0	**Introduction**
1.1	Purpose of meeting
1.2	Introduction of those present
2.0	**Confirmation of addenda letters issued**
3.0	**Responses to existing queries**
3.1	Contractor
3.2	Architect
3.3	Civil and structural engineer
3.4	Mechanical and electrical engineer
3.5	Other consultants
3.6	Quantity surveyors
3.7	Project manager
4.0	**Other additional information**
4.1	Contractor
4.2	Architect
4.3	Civil and structural engineer
4.4	Mechanical and electrical engineer
4.5	Other consultants
4.6	Quantity surveyors
4.7	Project manager
5.0	**Contractor's queries**
6.0	**Confirmation of tender arrangements**
6.1	Date
6.2	Time
6.3	Address
7.0	**Any other business**

APPENDIX B — Selection and appointment of contractors
B8 Returned tender review process

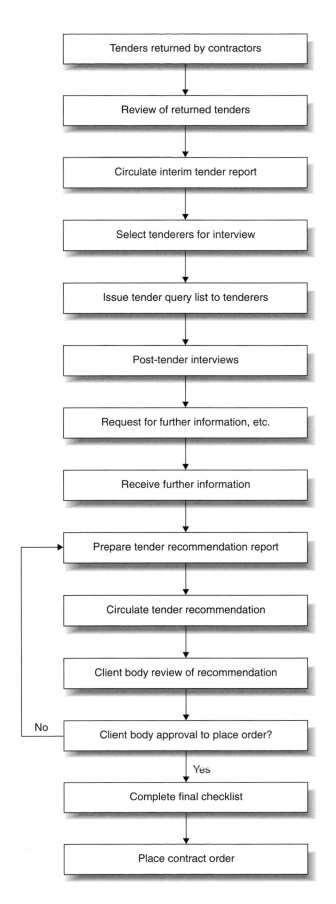

Action day	Comments	Checked by

APPENDIX B Selection and appointment of contractors
B9 Returned tender bids record sheet

Project Management Ltd			Form R1	
Ref. number:				
Contract title:				
Allocated budget: £		**Programme period:**		
No.	**Contractor qualifications, etc.**		**Prog.**	**Bid sum**
1.				
2.				
3.				
4.				
5.				
6.				
7.				
8.				

Signed by the undersigned as a true record of duly and properly

Received tender bids for _____

On this day _____ (day) _____ (month) _____ in the year _____

Signed:

Company:

Signed:

Company:

Signed:

Company:

Signed:

Company:

APPENDIX B   Selection and appointment of contractors
B10 Post-tender interview agenda

Project Management Ltd	Form A3

Ref. number:

Contract title:

1.0   Introduction

1.1   Introduction to those present

1.2   Purpose of meeting

2.0   Confirmation of contract scope and responsibilities

3.0   Detailed bid discussions

3.1   Contractual

3.2   Cost

3.3   Schedule

3.4   Method

3.5   Technical matters

3.6   Staffing labour and plant matters

3.7   Labour relations matters

3.8   Health and safety matters

3.9   Quality management

4.0   Contractor queries

5.0   Action and responses

5.1   Agreement of action items

5.2   Agreement of deadline dates for resolution of action items

APPENDIX B   Selection and appointment of contractors
B11 Final tender evaluation report

Project Management Ltd	Form R2

Ref. number:

Contract title:

1.0   Summary of final tender bids following post-tender interviews

2.0   Cost appraisal

3.0   Schedule appraisal

4.0   Method statement appraisal

5.0   Technical appraisal

6.0   Contractual appraisal

7.0   Quality management appraisal

8.0   Health and safety appraisal

9.0   Labour relations appraisal

10.0   Recommendation to place contract

Appendices

1   Completed tender bid of recommended tenderer

2   Addenda and other information issues during tendering

3   Mid-tender interview meeting minutes

4   Query lists and responses

5   Post-tender meeting minutes

6   Any other letters, etc. since tender issue

7   Summary of contract buy and any other items to be bought

APPENDIX B   Selection and appointment of contractors
B12 Approval to place contract order

Project Management Ltd	Form O1

Ref. number:

Contract title:

In accordance with clause _____ of the _____

_____ (Contract Form)

We _____

Do not have any objection to the placing of a contract with _____

for _____

All in accordance with the tender recommendation report submitted to us on the _____ (date).

In submitting the tender recommendation report the consultants are fully satisfied that the contractor has complied in full with the tender documents and is fully capable of carrying out the contract works.

Signed by: _____

Signed by: _____

Signed by: _____

APPENDIX B   Selection and appointment of contractors
B13 Final general checklist

Project Management Ltd	Form C3

Ref. number:

Contract title:

Check once again that the following were carried out:

☐   Long list

☐   Telephone selection questionnaires

☐   Contractor selection questionnaires, company accounts, references and any reports of visits to offices, factories and contracts

☐   Short list

☐   Pre-qualification interview minutes

☐   Tender list

☐   Substitute tender list

☐   Tender documents and checklist

☐   Tender query lists addendum letters prior to mid-tender interviews

☐   Mid-tender interview minutes

☐   Tender query lists addendum letters, etc. post mid-tender interviews

☐   Returned tender summary form and returned tender documents

☐   Interim tender analysis and recommendations report

☐   Post-tender query lists to contractors

☐   Post-tender interview minutes

☐   Post-tender addendum letters, etc.

☐   Final tender analysis and recommendations report

☐   Contractor acceptability final check

☐   Approval to place contract order

APPENDIX C — Engineering services commissioning checklist

Engineering services to be covered

■ Routinely:

- ○ water supply and sanitation
- ○ heating/cooling systems (boilers, calorifiers, chillers)
- ○ ventilation systems
- ○ air-conditioning
- ○ electrical (generators, switchboards, others)
- ○ mechanical (pumps, motors, others)
- ○ fire detection and protection systems
- ○ control systems (electrical, pneumatic, others)
- ○ telephone/communications

■ specialist

- ○ process plant for food, pharmaceutical, petrochemical or manufacturing activities
- ○ security (CCTV, sensors, access control)
- ○ facility management system
- ○ acoustic and vibration scans
- ○ lifts, escalators, others
- ○ IT systems, e.g. IBM, DEC, ICL.

Contract documents

■ Responsibilities – client/contractor/manufacturer:

- ○ bills of quantities/activity schedule items for commissioning activity with separate sums of clearly worded inclusion in M&E item descriptions

■ specification of commissioning:

- ○ provision for providing that commissioning is performed – observation test results
- ○ methods and procedures to be used, appropriate standards/codes of practice, e.g. CIBSE/IHVE/BSRIA/IEE/LPC/BS (see Appendix D).

■ provision for appropriate client access

■ client staff training

■ operating and maintenance manuals (as installed)

■ statutory approvals

■ record drawings and equipment software (as installed) and test certification

■ statutory approvals (lifts, fire protection, others)

■ insurance approvals.

Contractor's commissioning programme

■ Manufacturers' works testing

■ site tests prior to commissioning (component testing, e.g. a fan motor)

■ pre-commissioning checks (full system, e.g. air-conditioning, by contractor before demonstration to client)

- set to work (system by system)
- commissioning checks (including balancing/regulation)
- demonstration to client (system basis)
- performance testing (including integration of systems)
- post-commissioning checks (including environmental fine-tuning during facility occupancy).

APPENDIX D   Engineering services commissioning documents

CIBSE

Commissioning codes	A	Air distribution
	B	Boiler plant
	C	Automatic control
	R	Refrigerating systems
	W	Water distribution systems
	TM12	Emergency lighting

BSRIA

	TM 1/88	Commissioning HVAC systems divisions of responsibilities
	TN 1/90	European commissioning procedures
	AG 1/91	The commissioning of VAV systems in buildings
	AG 2/89	The commissioning of water systems in buildings
	AG 3/89	The commissioning of air systems in buildings
	AG 8/91	The commissioning and cleaning of water systems
	AH 2/92	Pre-commissioning of BEMS – a code of practice
	AH 3/93	Installation commissioning and maintenance of fire and security systems

HMSO

	HTM 17	Health building engineering installations commissioning and associated activities (hospitals)
	HTM 82	Fire safety in health care premises fire alarms and detection systems (hospitals)

Loss Prevention Council

	LPC	Rules for automatic sprinkler installations

IEE

Wiring regulations

British Standards

An extensive list of BS publications exists for specialised systems and equipment, e.g. gas flues, steam and water boilers, oil and gas burning equipment, electrical equipment, earthing machinery, etc.

IT

Cabling installation and planning guide of relevant 'equipment' manufacturer/supplier

APPENDIX E Handover checklists

Handover procedure
- architect's certificate
- Certificate of Practical Completion
- CDM health and safety file
- inspections and tests
- copies of certificates, approvals and licences
- release of retention monies
- final clean and removal of rubbish
- handover of spares
- meters read and fuel stocks noted.

Schedule
- remedial works
- Defects Liability Period and defect correction
- adjustment of building services
- client's fitting out.

Building owner's manual
- consultant's contributions
- format.

Operating and maintenance manuals, as-built drawings and C&T records
- servicing contracts established
- handover to facilities manager.

Letting or disposal
- schedule
- publicity
- strategy
- liaison
- documentation
- insurance.

Additional works
- contracts
- major service installations or adaptations
- fitting out
- shop fitting.

Final account, final inspection and Final Certificate arrangements

Liaison with tenants, purchaser or financier

Access by contractors
- remedial works
- additional contracts

Security
- key cabinet
- key schedule.

Publicity

Opening arrangements

Client's acceptance of building

Post-completion review/project close-out report

Inspection Certificates and Statutory Approvals

Fire officer inspections
- fire shutters
- fireman's lift
- smoke extract system/pressurisation
- foam inlet/dry riser
- fire dampers
- alarm systems
- alarm panels
- telephone link
- fire protection systems:
  - sprinklers
  - hose reels
  - hand appliances/blankets, etc.
- statutory signs.

Fire Certificate

Institution of Electrical Engineers' certificate

Water authority certificate of hardness of water

Insurer's inspections
- fire protection systems:
  - sprinkler
  - hose reels
  - hand appliances
- lifts/escalators
- mechanical services:
  - boilers
  - pressure vessels

○　electrical services

○　security installations

Officers of the court inspection (licensed premises)

Pest control specialists' inspection

Environmental health officer inspection

Building control officer inspection

Planning　　■　outline

　　　　　　　■　detailed including satisfaction of conditions

　　　　　　　■　listed building

Landlord's inspection

Health and safety officer's inspection

Crime prevention officer's inspection

APPENDIX F Client commissioning checklist

Brief Ensure roles and responsibilities for commissioning team are developed and understood progressively from the feasibility and strategy stages.

Budget schedule
- based upon a clear understanding and agreement of the client's objectives.

Commissioning action checklist
- investigate and identify commissioning requirements
- management control document.

Appointments
- commissioning team
- operating and maintenance personnel
- aftercare engineer
- job descriptions, time-scales and outputs must be documented and agreed.

Client operating procedures
- work practice standards
- health and safety at work requirements.

Training of staff
- services
- security
- maintenance
- procedures
- equipment.

Client equipment (including equipment rented for commissioning)
- schedule
- selection
- approval
- delivery
- installation.

Building services and equipment
- define/check standards required in tender specification
  - testing
  - balancing
  - adjusting } detail format of records
  - fine tuning
- marking and labelling, including preparation of record drawings
  - handover of spares } must be compatible with any planned maintenance or equipment standardisation policies
  - handover of tools

Maintenance
- acceptance by client's maintenance section from the client's construction and commissioning team
- arrangements
- procedures
- contracts.

Security	■ alarm systems
	■ telephone link
	■ staff routes
	■ access (including card access)
	■ fire routes
	■ bank cash dispensers.
Communications	■ telephones
	■ radios
	■ paging
	■ public address systems
	■ easy-to-read plan of building
	■ data links.
Signs and graphics	■ code of practice for the industry
	■ statutory notices – H&S, fire, Factories Act, unions.
Initiation of operations	■ final cleaning
	■ maintenance procedures (including manufacturers' specialist maintenance)
	■ cleaning and refuse collection
	■ insurance required by date and extent of cover will vary with the form of contract
	■ access and security (including staff identity cards)
	■ safety
	■ meter readings or commencement of accounts for gas, water, electricity, telephone and fuel oil
	■ equipping
	■ staff 'decanting'
	■ publicity
	■ opening arrangements.
Review operation of facility	■ at +6, 9 and 12 months (including energy costs)
	■ improvements and system fine-tuning
	■ defects reporting, correction and verification procedures
	■ latent defects.
Feedback	■ channelled through aftercare engineer if appointed.

Glossary

Throughout this work words in the masculine also mean the feminine and vice versa. Words in the plural include the singular, e.g. 'subcontractors' could mean just one subcontractor.

Aftercare engineer
The aftercare engineer provides a support service to the client/user during the initial 6–12 months of occupancy and is, therefore, most likely a member of the commissioning team.

As built drawings
Drawings provided for the building owner recording how the project was actually constructed.

Bill of quantities
Contract document listing items of work measured according to a Standard Method of Measurement, used in compiling the tender bids and as a basis for pricing variations.

Briefing
The process that enables the client to identify and agree with the project team the objectives, scope and detailed requirements for the project.

Change order
An alternative name for variation order, it indicates a change to the project brief.

Client
Owner and/or developer of the facility; in some cases the ultimate user.

Client adviser
An independent construction professional engaged by the client to give advice in the early stages of a project, as advocated by Latham.

Commissioning team
Client commissioning: predominantly the client's personnel assisted by the contractor and consultants. *Engineering services commissioning*: specialist contractors and equipment manufacturers monitored by the main contractor and consultants concerned.

Consultants
Advisers to the client and members of the project team. Also includes design team.

Contract administrator
The person who supervises the construction work on behalf of the client and who is empowered to issue instructions and certificates under the contract.

Contractor
Generally applied to (a) the main contractor responsible for the total construction and completion process; or (b) two or more contractors responsible under separate contractual provisions for major or high-technology parts of a very complex facility. (See **Subcontractor**.)

Design audit
Carried out by members of an *independent* design team providing confirmation or otherwise that the project design meets, in the best possible way, the client's brief and objectives.

Design freeze
Completion and client's final approval of the design and associated processes, i.e. no further changes are contemplated or accepted within the budget approved in the project brief.

Design team
Architects, engineers and technology specialists responsible for the conceptual design aspects and their development into drawings, specifications and instructions required for construction of the facility and associated processes.

Development surveyor	Provides information and advice on the environmental planning implications of the proposed facility, e.g. economic, social, financial and population trends.
Facilities management	Planning, organisation and managing physical assets and their related support services in a cost-effective way to give the optimum return on investment in both financial and quality terms.
Facility	All types of constructions, e.g. buildings, shopping malls, terminals, hospitals, hotels, sporting/leisure centres, industrial/processing/chemical plants and installations and other infrastructure projects.
Feasibility stage	Initial project development and planning carried out by assessing the client's objectives and providing advice and expertise in order to help the client define more precisely what is needed and how it can be achieved.
Handbook	See **Project handbook**.
Life-cycle costing	Establishes the present value of the total cost of an asset over its operating life, using discounted cash flow techniques, for the purpose of comparison with alternatives available. This enables investment options to be more effectively evaluated for decision making.
Occupation	Sometimes called *migration* or *decanting*. It is the actual process of physical movement (transfer) and placement of personnel (employees) into their new working environment of the facility.
Planning gain	A condition attached to a planning approval which brings benefits to the community at a developer's expense.
Planning supervisor	A consultant or contractor appointed by a client under the CDM Regulations to carry out this role.
Principal contractor	The contractor appointed by a client under the CDM Regulations to carry out this role.
Project brief	Defines the client's objectives, budget and functional requirements for the proposed facility.
Project handbook	Guide to the project team members in the performance of their duties, identifying their responsibilities and detailing the various activities and procedures (often called the project bible). Also called project execution plan, project manual and project quality plan.
Project team	Client, project manager, design team, consultants, contractors and subcontractors.
Risk factor	Associated with the anticipation and reduction of the effects of risk and problems by a proactive approach to project development and planning.
Schedule	Also called the project development schedule, it is an overall statement of the project approved plan and its stages, presented in a graphical form.
Strategy stage	During this stage a sound basis is created for the client on which decisions can be made allowing the project to proceed to completion. It provides a framework for the effective execution of the project.
Subcontractor	Contractor who undertakes specialist work within the project; known as specialist, works, trade, work package, and labour only.
Tenant	Facility user who is generally not the client or the developer.
User	The ultimate occupier of the facility.

Bibliography

The following is not intended to provide a comprehensive guide to the vast amount of literature available. Rather it is intended to support readers by directing them to supplementary titles which will allow construction project management and the intertwined processes to be evaluated and understood within the appropriate context.

- A Guide to Managing Health and Safety in Construction (1995), Health and Safety Executive

- A Guide to Project Team Partnering (2002), Construction Industry Council

- A Guide to Quality-based Selection of Consultants: A Key to Design Quality, Construction Industry Council

- Accelerating Change – Rethinking Construction (2002), Strategic Forum for Construction

- ACE Client Guide (2000), Association of Consulting Engineers

- Achieving Excellence through Health and Safety, Office of the Government Commerce

- Adding Value through the Project Management of CDM (2000), Royal Institute of British Architects

- Appointment of Consultants and Contractors, Office of the Government Commerce

- Association of Consulting Engineers (1991), Good Design is Good Investment. Advice to Client, Selection of Consulting Engineer, and Fee Competition.

- Benchmarking, Office of the Government Commerce

- Bennett, J (1985), Construction Project Management, Butterworths

- Best Value in Construction (2002), Royal Institution of Chartered Surveyors

- Briefing the Team (1996), Construction Industry Board

- British Property Federation (1983), Manual of the BPF System for Building Design and Construction

- British Standards Institution, Guide to Project Management BS 6079 – 1: (2000)

- Building a Better Quality of Life, A Strategy for More Sustainable Construction (2000), Department of Environment, Transport and the Regions, Health and Safety Executive

- Burke, R (2001), Project Management Planning and Control Techniques, 3rd edition

- Client Guide to the Appointment of a Quantity Surveyor (1992), Royal Institution of Chartered Surveyors

- Code of Estimating Practice, 5th edition (1983), The Chartered Institute of Building

- Code of Practice for Project Management for Construction and Development, 2nd edition (1996), The Chartered Institute of Building

- Code of Practice for Selection of Main Contractors (1997), Construction Industry Board

- Code of Practice for Selection of Subcontractors (1997), Construction Industry Board

- Constructing Success: Code of Practice for Clients of the Construction Industry (1997), Construction Industry Board

- Constructing the Team. Sir Michael Latham (1994), Final Report of the Government/Industry Review of Procurement and Contractual Arrangements in the UK Construction Industry (the Latham Report), HMSO

- Construction (Design and Management) Regulations 1994, Health and Safety Executive

- Construction (Design and Management) Regulations 1994, HMSO

- Construction (Health, Safety and Welfare) Regulations 1996, Health and Safety Executive

- Construction Best Practice Programme (CBPP) Fact Sheets

- Construction Health and Safety Checklist (Construction Information Sheet No. 17), Health and Safety Executive

- Construction Management Contract Agreement (Client/Construction Manager) (2002), Royal Institute of British Architects

- Construction Management Contract Guide (2002), Royal Institute of British Architects

- Construction Project Management Skills (2002), Construction Industry Council

- Control of Risk – A Guide to the Systemic Management of Risk from Construction (SP 125) (1996), Construction Industry Research and Information Association

- Cox, A and Ireland, P (forthcoming – 2003), Managing Construction Supply Chains, Thomas Telford

- Essential Requirements for Construction Procurement Guide, Office of the Government Commerce

- Essentials of Project Management (2001), Royal Institute of British Architects

- Facilities Management Contract, 2nd edition (2001), The Chartered Institute of Building

- Financial Aspects of Projects, Office of the Government Commerce

- Gray, C (1998), Value for Money, Thomas Telford

- Green D, editor (2000), Advancing Best Value in the Built Environment – A Guide to Best Practice, Thomas Telford

- Hamilton, A (2001), Managing Projects for Success, Thomas Telford

- Langford, D, Hancock, MR, Fellows, R and Gale, AW (1995), Human Resources Management in Construction, Longman

■ Lock, D (2001), Essentials of Project Management, Gower Publishing

■ Management Development in the Construction Industry – Guidelines for the Construction Professionals, 2nd edition (2001), published for the Institution of Civil Engineers by Thomas Telford Publishing

■ Managing Health and Safety in Construction. Construction (Design and Management) Regulations 1994. Approved Code of Practice and guidance. HSG224 HSE Books 2001, Health and Safety Executive

■ Managing Project Change – A Best Practice Guide (C556) (2001), Construction Industry Research and Information Association

■ Modernising Construction: Report by the Comptroller and Auditor General (2001), HMSO

■ Modernising Procurement: Report by the Comptroller and Auditor General (1999), HMSO

■ Morris, PWG (1998), The Management of Projects, Thomas Telford

■ Murdoch, I and Hughes, W (2000), Construction Contracts: Law and Management, E & FN Spon

■ Partnering in the Public Sector – A Toolkit for the Implementation of Post-award, Project Specific Partnering on Construction Projects (1997), European Construction Institute

■ Partnering in the Team (1997), Construction Industry Board

■ Planning: Delivering a Fundamental Change (2000), Department of Environment, Transport and the Regions

■ Potts, K (1995), Major Construction Works: Contractual and Financial Management, Longman

■ Procurement Strategies, Office of the Government Commerce

■ Project Evaluation and Feedback, Office of the Government Commerce

■ Project Management (2000), Royal Institute of British Architects

■ Project Management Body of Knowledge (2000), Association for Project Management

■ Project Management in Building, 2nd edition (1988), The Chartered Institute of Building

■ Project Management Memorandum of Agreement and Conditions of Engagement, Project Management Panel, RICS Books

■ Project Management Planning and Control Techniques (2001), Royal Institute of British Architects

■ Project Management Skills in the Construction Industry (1996), Construction Industry Council

■ Quality Assurance in the Building Process (1989), The Chartered Institute of Building

■ Rethinking Construction – Report of the Construction Task Force to the Deputy Prime Minister on the Scope for Improving the Quality and Efficiency of UK Construction (the Egan Report) (1998), Department of Environment, Transport and the Regions

- Risk Analysis and Management for Projects (1998), Institution of Civil Engineers and Institute of Actuaries

- Safety in Excavations (Construction Information Sheet No. 8), Health and Safety Executive

- Selecting Consultants for the Team (1996), Construction Industry Board

- Selecting Contractors by Value (SP 150) (1998), Construction Industry Research and Information Association

- Teamworking, Partnering and Incentives, Office of the Government Commerce

- The Procurement of Professional Services: Guidelines for the Application of Competitive Tendering (1993), Thomas Telford on behalf of CIC

- The Procurement of Professional Services: Guidelines for the Value Assessment of Competitive Tenders (1994), Construction Industry Council

- Thinking about Building? Independent Advice for Small and Occasional Clients, Confederation of Construction Clients

- Thompson, P and Perry, JG (1992), Engineering Construction Risks – A Guide to Project Risk Analysis and Risk Management, Thomas Telford

- Turner, JR (1999), The Handbook of Project-based Management, McGraw-Hill

- Value by Competition (SP 117) (1994), Construction Industry Research and Information Association

- Value for Money in Construction Procurement, Office of the Government Commerce

- Value Management in Construction: A Client's Guide (SP 129) (1996), Construction Industry Research and Information Association

- Walker, A (2002), Project Management in Construction, Blackwell

- Whole Life Costs, Office of the Government Commerce

USEFUL WEBSITES

- Construction Industry Research and Information Association – www.ciria.org.uk

- Health and Safety Executive – www.hse.gov.uk

- Her Majesty's Stationery Office – www.hmso.gov.uk

- Institution of Civil Engineers – www.ice.org.uk

- Movement For Innovation – www.m4i.org.uk

- National Audit Office – www.nao.gov.uk

- Rethinking Construction – www.rethinkingconstruction.org

- Royal Institute of British Architects – www.architecture.com

- Royal Institution of Chartered Surveyors – www.rics.org.uk

- The Chartered Institute of Building – www.ciob.org.uk

- The Commission for Architecture and the Built Environment – www.cabe.org.uk

- The Confederation of Construction Clients – www.clientsuccess.org.uk
- The Construction Best Practice Programme – www.cbpp.org.uk
- The Construction Industry Council – www.cic.org.uk
- The Construction Industry Training Board – www.citb.org.uk
- The Housing Forum – www.thehousingforum.org.uk
- The Local Government Task Force – www.lgtf.org.uk
- The Office of Government Commerce – www.ogc.gov.uk

Index